E-GOVERNANCE IN EUROPEAN AND
SOUTH AFRICAN CITIES

The European Institute for Comparative Urban Research, EURICUR, was founded in 1988 and has its seat with Erasmus University Rotterdam. EURICUR is the heart and pulse of an extensive network of European cities and universities. EURICUR's principal objective is to stimulate fundamental international comparative research into matters that are of interest to cities. To that end, EURICUR coordinates, initiates and carries out studies of subjects of strategic value for urban management today and in the future. Through its network EURICUR has privileged access to crucial information regarding urban development in Europe and North America and to key persons at all levels, working in different public and private organizations active in metropolitan areas. EURICUR closely cooperates with the Eurocities Association, representing more than 100 large European cities.

As a scientific institution, one of EURICUR's core activities is to respond to the increasing need for information that broadens and deepens the insight into the complex process of urban development, among others by disseminating the results of its investigations by international book publications. These publications are especially valuable for city governments, supranational, national and regional authorities, chambers of commerce, real estate developers and investors, academics and students, and others with an interest in urban affairs.

EURICUR website: http://www.euricur.nl

This book is one of a series to be published by Ashgate under the auspices of EURICUR, the European Institute for Comparative Urban Research, Erasmus University Rotterdam. Titles in the series are:

Growth Clusters in European Metropolitan Cities
Leo van den Berg, Erik Braun and Willem van Winden

Information and Communications Technology as Potential Catalyst for Sustainable Urban Development
Leo van den Berg and Willem van Winden

Sports and City Marketing in European Cities
Leo van den Berg, Erik Braun and Alexander H.J. Otgaar

Social Challenges and Organising Capacity in Cities
Leo van den Berg, Jan van der Meer and Peter M.J. Pol

City and Enterprise
Leo van den Berg, Erik Braun and Alexander H.J. Otgaar

The Student City
Leo van den Berg and Antonio P. Russo

European Cities in the Knowledge Economy
Leo van den Berg, Peter M.J. Pol, Willem van Winden and Paulus Woets

Urban Issues and Urban Policies in the New EU Countries
Ronald van Kempen, Marcel Vermeuten and Ad Baan

The Safe City
Leo van den Berg, Peter M.J. Pol, Giuliano Mingardo and Carolien J.M. Speller

E-Governance in European and South African Cities

The Cases of Barcelona, Cape Town, Eindhoven, Johannesburg, Manchester, Tampere, The Hague and Venice

LEO VAN DEN BERG
ANDRE VAN DER MEER
WILLEM VAN WINDEN
PAULUS WOETS

European Institute for Comparative Urban Research
Erasmus University Rotterdam
The Netherlands
www.euricur.nl

Routledge
Taylor & Francis Group

LONDON AND NEW YORK

First published 2006 by Ashgate Publishing

Reissued 2018 by Routledge
2 Park Square, Milton Park, Abingdon, Oxon OX14 4RN
605 Third Avenue, New York, NY 10017

First issued in paperback 2021

Routledge is an imprint of the Taylor & Francis Group, an informa business

A Library of Congress record exists under LC control number: 2006025016

Notice:
Product or corporate names may be trademarks or registered trademarks, and are used only for identification and explanation without intent to infringe.

Publisher's Note
The publisher has gone to great lengths to ensure the quality of this reprint but points out that some imperfections in the original copies may be apparent.

Disclaimer
The publisher has made every effort to trace copyright holders and welcomes correspondence from those they have been unable to contact.

ISBN 13: 978-0-815-38875-3 (hbk)
ISBN 13: 978-1-351-15916-6 (ebk)
ISBN 13: 978-1-138-35685-6 (pbk)

DOI: 10.4324/9781351159166

Contents

List of Figures

Figures

List of Tables

Preface

In 1999, Euricur conducted an explorative study on urban ICT policies in Europe. By that time, many cities had become very active in this field. The study described how ICTs could contribute to sustainable urban development, and what cities were actually doing to benefit most from the new possibilities. This was one of the first European comparative analyses in this field, going beyond merely comparing basic indicators.

After completion of the project we concluded that there was much more to the topic, and so we launched a second study that would focus more on E-governance issues. In this underlying study, we make a distinction between various types of ICT policies (e-access, e-infrastructure and e-content), and pay ample attention to the organizational issues that are associated with the implementation of ICT projects. Also, the project's scope reaches beyond Europe, and now also includes two South African cities.

This report could not have been produced without the cooperation of a number of people and institutions. We would like to thank the contact persons in the participating cities: Lluis Olivella (Barcelona), Nirvesh Sooful (Cape Town), Peter de Wit (Eindhoven), Zamile Mazantsana (Johannesburg), Dave Carter (Manchester), Jarmo Viteli (Tampere), Marten Buschman (The Hague), and Jan van der Borg and Andrea di Mercato (Venice). Furthermore, we are grateful for the financial contributions and support of IBM and Cisco, and Antonio Paolo Russo for writing the chapter on Venice. Finally, we thank our secretary Ankimon Vernède for her support and kindness.

Rotterdam, January 2005

Chapter 1

Introduction

The information society entails a number of fundamental changes, many of which manifest themselves in cities. These changes affect urban structures (Castells, 1996; Hall, 1998), forms (van den Berg, 1987; Mitchell, 1999), economies (Hall, 1998; Storper, 1996; Thrift, 1996) and societies (Castells, 1996; Sassen, 2001). The development and application of ICTs (information and communication technologies) lies at the heart of these transformations. ICTs can be described as the melting of computer technology, telecommunications, electronics and media (van Rijsselt and Weijers, 1997). Examples of new ICTs are the personal computer, but also the Internet, mobile telephone, cable television and electronic payment systems are included. In the last decade, innovations in communications and information technology have been introduced at rapid speed (Castells, 1996; Forrester, 1993).

There is a growing literature about the way ICTs are changing cities. In this book, we want to contribute to this debate. In our approach, we intend to move away from the abstract macro-idea of 'the information society' and instead stress diversity and the 'local colour' of the information society, on the urban level. This approach includes the newer strands of technology research in social sciences that focus on the context-dependency of the uptake of technologies. New technologies do not fall out of the sky into a homogeneous landscape and then change it completely: their development and application is embedded in existing economic, institutional, social and spatial structures, and changes them in rather subtle ways.

The focus in this book is on Internet-related technologies and services, as they are relatively new and have the profoundest social and economic implications. Our starting point is that, for a number of reasons, the manifestation of the 'information society' varies considerably from city to city. To reveal this diversity, we developed a conceptual framework that helps us to unravel the local colour of the information society in cities. We make a distinction between three manifestations of the information revolution: local electronic content, local access to new technologies and local electronic infrastructure. We suggest that the interaction between the three manifestations drives the dynamics of the local information society. We also suggest how policy – on several levels – might influence these dynamics.

The second part of the book is about ICT policy in a number of cities. The case studies show major differences in policy orientation, reflecting different policy priorities. They also reveal the importance of the national economic, political and legal contexts as determinants of the shape of the local information society. In this book, we explicitly address the role of private ICT companies, with a focus on the impact of investment decisions of telecom firms on the local endowment of electronic infrastructure.

This book is based on an international comparative study into 'e-governance' strategies. In eight cities we have studied local ICT policies. Our case studies were Barcelona (Spain), Cape Town and Johannesburg (South Africa), Eindhoven and The Hague (The Netherlands), Manchester (United Kingdom), Tampere (Finland) and Venice (Italy). For our purposes (showing and analysing the variety of local manifestations and policies of the information society), this is a good sample of cities. They are located in different countries, which may reveal the importance of the national context. They differ considerably in economic structure and performance. Some are very specialized, albeit in different sectors (e.g. Venice in tourism, The Hague in administrative functions), others have a more diversified economy. As could be expected, each of the cities has its particular focus in ICT policy. However, all the cities share a relatively high ambition level and high expectations of ICT policies.

For each of the cities, we started with an analysis of the available information on the local ICT situation and official 'e-strategy' documents. After that, in each of the cities we have interviewed a number of government officials responsible for the local ICT policy as well as private companies that are involved in the implementation of the policies. Also, we have interviewed ordinary citizens to hear their opinion as 'policy receivers'.

Organization of the Book

This book is organized as follows. In Chapter 2, we present a conceptual framework in which to analyse the local manifestation of the information society. Also, the concept of e-governance is introduced and elaborated. The subsequent chapters (3–10) describe and analyse local e-governance practices in our case cities. Finally, Chapter 11 synthesises the findings, compares the case studies and draws conclusions.

References

Castells, M. (1996), *The Rise of the Network Society. The Information Age: Economy, Society and Culture*, vol. 1, Blackwell, Oxford.

Forrester, T. (1993) *Silicon Samurai: How Japan Conquered the world Information Technology Industry*, Blackwell, Oxford.

Hall, P. (1998), *Cities in Civilization: Culture, Innovation and Urban Order*, Weidenfeld and Nicolson, London.

Mitchell, W.J. (1999), 'Equitable Access to the Online World, in High Technology and Low-income Communities', in D.A.B. Schön, A.B. Sanyal and W.J. Mitchell (eds), *Prospects for the Positive Use of Advanced Information Technology*, MIT Press, Cambridge.

Sassen, S. (2001), 'Debates and Developments – Impact of Information Technologies on Urban Economies and Politics', *International Journal of Urban and Regional Research* 25 (2), pp. 411–18.

Storper, M. (1996), 'The World of the City: Local Relations in a Global Economy', mimeo, School of Public Policy and Social Research, University of California.

Thrift, N. (1996), 'New Urban Eras and Old Technological Fears: Reconfiguring the Goodwill of Electronic Things', *Urban Studies* 33 (8), pp. 1463–93.

Van den Berg, L. (1987), *Urban Systems in a Dynamic Society*, Aldershot, Gower.

Van Rijsselt, R.T.J. and Weijers, T.C.M. (1997), 'Ouderen en de informatiesamenleving. Een verkenning van opvattingen over aansluiting en uitsluiting', Werkdocument 60, Rathenau Instituut.

Chapter 2

Framework of Reference

1   Introduction

This chapter introduces a framework of reference with the help of which the cases will be analysed. The framework is based on a literature review. In the framework we make a distinction between three types of 'footprint' of ICTs in urban areas: access (elaborated in section 2); content (section 3); and infrastructure (section 4). These 'footprints' interact with each other (section 5). Section 6 discusses the role of local governments and introduces the concept of e-governance.

2   Electronic Access

The first and most basic manifestation concerns the degree of access to technologies by the (various segments of) the urban population. Access to ICT has several dimensions. It includes not only the ownership of hardware devices, but also the capabilities to use information technologies, and access to the Internet (SCP, 2000; Mitchell, 1999a). On several geographical levels, we can witness varying degrees of access to new technologies. On a global scale, there is a digital divide between the developed world and the developing world (Vlam and Westra, 2002). Within countries (both developing and developed), there are substantial differences between large cities and rural areas, but also among large cities (DTI, 2000). Within cities, finally, there are large differences between districts (Graham, 2000). There is now a rich and growing literature on the determinants of access to technology (see van den Berg and van Winden, 2002, for an overview). Most accounts point at education levels and income as key factors. SCP (2000) found that the adoption of PCs and the Internet was positively related to cognitive, social and material resources of individuals.

3   Electronic Infrastructure

The second manifestation of the information society in cities is the electronic infrastructure. The various types of infrastructures (copper, coaxial lines, wireless networks, fibre-optic lines) can be regarded as the transportation system carrying the bits and bytes of the information society. The infrastructure landscape in cities has changed dramatically in the last decade. Most notably, the number of electronic infrastructure networks has increased (several new mobile networks have been put in place in the last decade, but also high-bandwidth fixed lines and satellite-based systems). Second, the spatial differences in infrastructure endowment have become wider, due to telecom markets liberalization and a declining prevalence of universal service obligations. The quality and availability of electronic infrastructure differs both within and between cities. Typically, because of market size, larger cities are better endowed than smaller cities or rural areas and within cities richer neighbourhoods and business districts have better infrastructures than poor neighbourhoods. In this perspective, Graham (1998) notes the emergence of premium network spaces. These are very localized areas in large cities (like London's financial district) that have superior connections both internally and also with similar places in other cities. For the location of business, particularly information-intensive service companies, the quality of broadband access is a major location factor (Healey and Baker, 2001). Broadband is different from 'narrowband' dial-up access in two important respects: first, it offers more capacity, and second, most broadband technologies entail an 'always on' connection: the user does not need to dial into a network, but is always online.

Table 2.1   Speed of connection of different modes

	Download	**Upload**
ISDN	One way: 128 Kbps; Both ways: 64 Kbps	One way: 128 Kbps; Both ways: 64 Kbps
Power line	1 Mbps–2 Mbps	1 Mbps–2 Mbps
DSL	6-8 Mbps max	640 Kbps
Cable	27 Mbps	2.5 Mbps
Fibre optic	50 Mbps–20 Gbps	50 Mbps–20 Gbps

Source: BDRC (2001).

4   Electronic Content

As a third manifestation, we discern the quality and availability of local electronic content. What is local content? We define it as electronically available information, interactive services or other web content related to or concerned with a specific locality. Examples of local content are the local newspaper on the Internet, websites on the traffic situation in the city, information about events in the city, or the electronic services that the local administration offers to its citizens. It also includes the websites of firms or institutes that primarily serve a local market, such as community organizations, education institutes and non-profit organizations. Finally, it includes local virtual communities, such as self-help groups, newsgroups etc. As we will show in our final synthesis chapter, our case cities differ widely in the quality and quantity of local content. City administrations play a large role in the determination of the quantity and quality of local content, as they are one of the most important 'suppliers' of it. But also, much depends on the local 'organizing capacity' of individual sectors to use the Internet as a new medium to communicate with clients or as a marketing tool.

5 Interaction and Dynamics

Cities are different in terms of content, access and infrastructure. Nevertheless, there are commonalities in the way each of the three 'manifestations' have an impact on each other. There are strong indications that the three local manifestations of the information society are interdependent and sometimes mutually reinforcing (see Figure 2.1). We suggest that its dynamics can be represented as a local 'digital flywheel', which functions as follows. If there are more ICT users (access) in a city, it becomes more interesting for companies or any other actors to develop new services (content). For instance, online grocers normally start their activities in areas where Internet penetration is highest. Alternatively, more (or better) electronic services (content) may increase the number of local users. If there are better online products or services available, the Internet becomes more useful and more people are likely use it. This interdependence between access and content is well known in the economic literature on technology adoption (see Leighton, 2001). In many instances, a 'killer application' can speed up the adoption of a new technology very rapidly. The quality of local content is probably not the key factor for an individual's decisions to buy a computer and go on the Internet. Nevertheless, several studies suggest that local information and services are very important for citizens (Anttiroiko, 1998; Servon and Nelson, 2001; Baines, 2002).

The quality of the local electronic infrastructure is linked to both access and content. On the one hand, higher levels of access and more electronic services will increase the demand for bandwidth and make high-level (broadband) electronic infrastructure more profitable. Telecom companies are more likely to offer high bandwidth services in areas where demand is greatest. Alternatively, if the quality of the local infrastructure is upgraded, this will evoke improved e-services (those which require broadband) and again attract more local users.

For cities, 'turning the flywheel on' may bring benefits in several respects. Improved electronic services mean a higher quality of life for inhabitants: they have better access to improved amenities. E-government services may save public spending and reduce local taxes to the benefit of citizens and/or firms. With the public sector responsible for up to 50 per cent of GDP in many Western European countries, the potential cost savings are clearly substantial. The quality of local electronic infrastructure is a factor of growing importance to attract or retain inhabitants (Healey and Baker, 2001). Wired homes have the potential for being seen as more upmarket and desirable than others (Baines, 2002). Virtual communities can contribute to safety, social cohesion and political participation

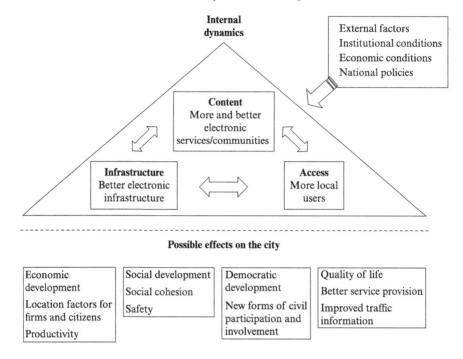

Figure 2.1   The 'digital flywheel'

(van Winden, 2001). High-quality infrastructure is also important to attract or retain firms in the region (Healey and Baker, 2001). Furthermore, policies may bring 'first mover advantages'. If a region manages to create early mass in users and infrastructure, local firms may build an innovative edge. In particular, early critical mass of users may attract innovative companies and people into the city. The system takes off when a critical mass of users is reached.

A question that comes to mind is how local is the 'local flywheel' really? Clearly, its engine is not solely fuelled by local factors. External factors play an important role, too. In the first place, national institutional conditions matter. Our South African case study suggests that it makes a big difference whether the telecom market is liberalized and competitive or not. All other kinds of legislation influence the flywheel as well, for instance, electronic privacy and security legislation. Second, general economic conditions play a role. ICT use is strongly related to economic development levels. Richer countries and cities tend to have higher levels of access, more content to offer and a higher quality of infrastructure. Third, national policies can strongly influence the different parts of the flywheel. Regarding access, many countries have nationwide programmes for ICT in education or access policies for disadvantaged communities. In the field of content, national policies may encourage cities or other public entities to develop e-strategies, and thus speed up the quantity and quality of content offered.

Despite all this, our study has revealed that there is still sufficient scope for urban policy makers to do something.

6 The Role of Local Governments: The Concept of E-Governance

Public policy with regard to ICT is often referred to as e-government. Silcock (2001) defines e-government as the use of technology by governments to enhance the access to and delivery of government services to benefit citizens, business partners and employees. The term 'e-government' is strongly associated with administrative governing by a single actor. However, in this book we are not only interested in the uptake of ICTs by local government itself, but just as much in the role of local government in the processes of uptake of ICTs by local population, communities and businesses and its role in infrastructure provision. It is clear that in these processes local government competencies are less direct, and depend critically on cooperation with other actors, such as IT companies, local communities and local business sectors. It has to operate in flexible networks to get things done.

In this light, the concept of governance developed in the institutional economic literature is useful for our purposes. It puts less emphasis on the directive capacities of local government, and more on its ability to engage in networks with other organizations. Mistri (1999) defines governance as the capacity of local administrations to guide the growth process in a dialectic exchange with social organizations and firms. Jessop (1997) describes governance as 'the complex art of steering multiple agencies, institutions and systems which are both operationally autonomous from another and structurally coupled through various forms of reciprocal interdependence' (p. 95).

In this spirit, we introduce the term 'e-governance', which can be described as the capacity of local administrations, in a dialectic exchange with social organizations, citizens and firms, to deploy information and communications technologies to achieve urban policy goals. An important element of this description is the representation of ICTs not as a means in themselves but as instrumental for the achievement of policy goals. In line with our distinctions in the previous section, we discern three dimensions of e-governance. First, local content governance is the capacity of local administrations to provide, create or promote user-friendly Internet or other electronic content related to a specific locality. Second, local access governance is the capacity of local administrations to provide access to new ICTs for the urban population at large. Third, local infrastructure governance is the capacity of local administrations to influence the provision and spatial distribution of electronic infrastructure (copper, coax cable, fibre, and eventually other technologies).

In different contexts, policy goals and orientations can be very different. This will have consequences for the form and focus of e-governance. Some local governments may design e-strategies primarily to promote social inclusion or fight 'digital divides' while others will target economic growth and development. Also, for political (or other) reasons, cities differ in their degree of interventionism in technology adoption processes. 'Interventionist' urban leaders are more likely to see a role for public policies to counter undesirable market outcomes or promote more equity than 'laissez-faire' local governments.

In our e-governance concept, the composition and quality of local networks deserves generous attention. Each dimension of local e-governance involves several partners or stakeholders: typically, these are local governments, citizens and technology suppliers, but also other parties may be involved such as other public agencies (municipal departments, financial service companies or other content providers). The different dimensions of e-governance (content, access and infrastructure) may require different approaches. In our framework, networks are not only a means to get things done with urban stakeholders, but can also be a powerful tool to influence the 'external factors' that influence the digital flywheel.

In each of the cities we have checked how the private sector is involved in policies, how bottom-up initiatives in the city are aligned with general visions and strategies on the urban level, and how public agencies (within the city, but also on the national and international levels) cooperate in various ICT-related policy fields.

In the following eight chapters, for each of the case studies we describe the e-strategy of the city and further elaborate each city's content, access and infrastructure policies. We also link the ICT policy to the specific urban social, economic and spatial context, and discuss the contribution of the ICT policies to the challenges and threats that the individual cities face.

References

Anttiroiko, A. (1998), 'Planting the Seeds of Economic Growth and Social Welfare: Local and Regional Governments in Finland and Korea Facing the Challenge of the Information Age', paper prepared for the International Conference on Electronic Democracy EDI, Korea.

Baines, S. (2002), 'Wired Cities', *Communications International*, April, pp. 21–25.

DTI (2000), 'Closing the Digital Divide: Information and Communication Technologies in Deprived Areas', a report by Policy Action Team 15, Department of Trade and Industry, London.

Graham, S. (1998), 'The End of Geography or the Explosion of Space? Conceptualising Space, Place and Information Technology', *Progress in Human Geography* 2, pp. 165–85.

Graham, S. (2000), 'Constructing Premium Network Spaces: Reflections on Infrastructure Networks and Contemporary Urban Development', *International Journal of Urban and Regional Research* 24 (1), pp. 183–200.

Healey and Baker Consultants (2001), *European E-locations Monitor*.

Jessop, B. (1997), 'Capitalism and its Future: Remarks on Regulation, Government and Governance', *Review of International Political Economy* 4 (3), pp. 561–81.

Leighton, W.A. (2001), *Broadband Deployment and the Digital Divide: A Primer*, Policy Analysis No. 410, August 7.

Mistri, M. (1999), 'Industrial Districts and Local Governance in the Italian Experience', *Human Systems Management* 18 (2), pp. 131–39.

Mitchell, W.J. (1999a), *E-topia: Urban Life Jim, But Not as We Know It*, MIT Press, Cambridge.

Mitchell, W.J. (1999b), 'Equitable Access to the Online World, in High Technology and Low-income Communities', in D.A.B. Schön, A.B. Sanyal and W.J. Mitchell (eds), *Prospects for the Positive Use of Advanced Information Technology*, MIT Press, Cambridge.

Servon, L.J. and Nelson, M.K. (2001), 'Community Technology Centers and the Urban Technology Gap', *International Journal of Urban and Regional Research* 25 (2), pp. 419–26.

SCP (2000), *Digitalisering van de leefwereld: een onderzoek naar informatie- en communicatietechnologie en sociale ongelijkheid*, Sociaal en Cultureel Planbureau, Den Haag.

Silcock, R. (2001), 'What is E-government?', *Parliamentary Affairs* 54, pp. 88–101.

van den Berg, L. (1987), *Urban Systems in a Dynamic Society*, Aldershot, Gower.

van den Berg, L. and van Winden, W. (2002), 'Should Cities Help their Citizens to Adopt ICTs? On ICT Adoption Policies in European Cities', *Environment and Planning C* 20 (2), pp. 263–79.

van Winden, W. (2001), 'The End of Social Exclusion? On Information Technology Policy as a Key to Social Inclusion in Large European Cities', *Regional Studies* 35 (9), pp. 861–77.

Vlam, P. and Westra, A. (2002), 'Waterputten of Internetten?', *Computable* 20, pp. 52–57.

Chapter 3

The Case of Barcelona

1 Introduction

In this chapter we will describe and analyse Barcelona's e-governance strategies. We will start in section 2 with an overview of the city. Section 3 offers a general context description of the challenges that the city of Barcelona faces. This puts the city's e-governance efforts into perspective. In section 4, we summarize the city's ICT strategy. In sections 5, 6 and 7 we analyse the issues of content, access and infrastructure respectively. Section 8 concludes.

2 Barcelona: Profile of the City

The metropolitan area of Barcelona has 4.5m inhabitants. Although it was abolished as an institutional entity in 1985, it can be considered as a functional urban region with a surface area of 3,200km^2.

The geographical area known as the metropolitan area includes two conurbations surrounding Barcelona: the Metropolitan Environmental Authority and the Metropolitan Region. The Metropolitan Environmental Authority is closest to the city. It is made up of Barcelona and 32 surrounding municipalities. The Metropolitan Region is more extensive and laid out within the Territorial Plan of Catalonia. It is made up of the following counties: Barcelonès (six municipalities, including Barcelona), Alt Penedès, Baix Llobregat, Garraf, Maresme, Vallès Occidental and Vallès Oriental.

The wider region of Catalonia is split up into four provinces with 767 municipalities altogether. The size of the municipalities ranges from as little as 25 to as many as 1.5m inhabitants.

According to the official registers the central municipality of Barcelona has 1.5m inhabitants. It is the second largest city in Spain. However, many people that live in the city are not registered here. For fiscal or other reasons they prefer to be registered in another city, where they might have a second house.

In 1984, due in part to the social and infrastructural changes in the city during the 1960s and 1970s, the current territorial division of the city of Barcelona into city districts was established. This was the culmination of a drawing-up process carried out by a broad-based commission in which municipal authorities worked alongside a number of institutions (Federation of Neighbourhood Associations, professional associations, Chamber of Commerce, Department of Public Works,

Figure 3.1 Barcelona Metropolitan Region

Source: Barcelona Department of Statistics.

Table 3.1 Population of Barcelona 1 January 2001

	Population	Surface (km²)	Density (inhab./km²)
Barcelona	1,527,190	101.0	15,120.7
Metropolitan Region	4,482,623	3.235,6	1,357
% Barcelona/Metropolitan Region	34.0	3.1	–
Catalonia	6,506,440	31,895.3	204.0
% Barcelona/Catalonia	23.5	0.3	–
Spain	41,837,894	506,030.0	82.7
% Barcelona/Spain	3.7	0.02	–

Source: Barcelona Statistical Department.

etc.) and local experts. Figure 3.2 shows the classification of the population according to educational level.

The modern parts of the city of Barcelona are the result of developments that started with the Olympic games in 1992. Many infrastructures were built at that time, such as the city ring roads, fibre-optic networks, the Olympic village, the

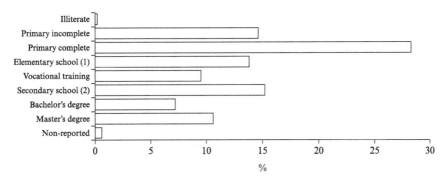

Figure 3.2 Educational level of the population in Barcelona

Source: Barcelona Statistical Department.

new port and many utilities. These developments continued over the 1990s and are still going on. A good example is the redevelopment of an old industrial area (Poblenou) into an area for high-tech industries, institutions for higher education, modern housing and a mixture of amenities.

Poblenou used to be the most important productive area of Catalonia. However, its industries moved to more peripheral areas. The regeneration plan for Poblenou (22@BCN) encompasses the transformation of this obsolete industrial area into spaces that are related preferably to new economy business initiatives. The 22@BCN project represents the most important urban transformation undertaken in the city in recent years and is possibly the last on this scale. This is because Poblenou is the last big city area that, due to its low urban density, can embark on a project of this nature. The infrastructural investments in Poblenou amount up to three times those made for the Olympic Games in 1992.

Poblenou will be characterized by emerging activities related to new information and communications technologies (ICT), research, design, publishing, culture, multimedia activities and knowledge and database management.

The economic profile of Barcelona shows that about half of the enterprises are active in the field of services (see Table 3.2).

3 Barcelona's Municipal Policy for the Twenty-first Century

The first strategic plan for Barcelona Metropolitan Region (published in 1990) aimed to consolidate Barcelona as a dynamic European metropolis, influential on the macro region in which it is located, with a modern quality of life, socially balanced and firmly rooted in the Mediterranean culture. The second strategic plan (1994) accentuated the integration of the Barcelona area into the international economy. According to the third strategic plan (1999) the Metropolitan Region of Barcelona should consolidate its position as one of the most important metropolitan regions in the European city network and must contribute with the purpose of connecting that

Table 3.2   Economic activities in Barcelona

Economic activities	1996		2001	
	N	%	N	%
Enterprises	142,588	100.0	156,110	100.0
Agriculture	11	0.0	7	0.0
Industry	14,932	10.5	13,962	8.9
Construction	9,134	6.4	12,370	7.9
Wholesale	13,656	9.6	13,935	8.9
Retail	41,021	28.8	40,653	26.0
Services	63,834	44.8	75,183	48.2
Professionals	41,332		45,182	

Source: Barcelona Statistical Department

network to the broader network of cities of the world through its own particularities
and identity. The third plan contains five strategic action lines:

1 To continue *positioning the Barcelona Metropolitan Region* as one of the most
   active and sustainable urban areas of the European Union.
2 To prioritize policies that *stimulate an increase in employment*. This should
   be done with regard to less favoured groups in particular, i.e. women, young
   people and those over 45 years old. This policy should be complemented by
   others that also stimulate an increase in the activity rate.
3 To facilitate the *evolution towards a city of knowledge* promoting new sectors
   of activity, within a framework of efficiency and participation, in order to
   ensure a stable quality of life and progress for its citizens.
4 To *ensure the social cohesion* of the citizens by deepening the participative
   culture of the city and creating the necessary spaces for participation.
5 To have *a significant role in the growth of Europe* and to develop a specific
   position, within Spain and especially abroad, with the Mediterranean and
   with Latin America. This is conceived as a factor which will greatly increase
   the city's internal attraction.

From the municipality of Barcelona's perspective these action lines contribute
to different profiles of the city.

The Connected City

Imperfect infrastructures (airport, roads, rail, telecommunications) appear to
be obstacles to attaining the mission assigned in the third strategic plan. The
measures concerned with infrastructures are distributed amongst three basic

lines: the Barcelona Metropolitan Region, the international profile, and the city of knowledge. Telecommunications are also identified as one of the key obstacles to be overcome.

The Open Enterprising City

Education, training and research are the great challenges to Barcelona in order to attain a good level of employment and social inclusion. An enterprising spirit in its citizens is conceived as an indispensable condition to ensuring the dynamism necessary to continually open up to new economic activities.

A Region of Cities

The Barcelona Metropolitan Region constitutes a compact group of cities. There is potential for coordinated services and significant added value that could be used to exploit the opportunities that the global economy offers to this type of urban agglomeration.

A City for People

Barcelona wants to be a city that attempts to ensure a high quality of life for all its citizens, especially with regard to employment, culture and mobility.

National and Regional Context

The administration of Spain is divided into three levels: state, regional and local. The local level includes both provinces and municipalities. The prime role of the provinces is service oriented. They support the smaller communities in the region (e.g. maintaining the citizens' register), and run the libraries.

Barcelona is the capital of the autonomous region of Catalonia, which has 6m inhabitants. In and around Barcelona the *national identity* is experienced at the level of the Catalonia region.

Barcelona is a bilingual city. Catalan and Castellan (Spanish) are the official languages. The municipal websites are presented in both languages. The municipal IT department has developed an automated translation tool for documents.

Education and healthcare are primarily responsibilities at regional level, although – for historical reasons – Barcelona has some municipal schools and hospitals of its own.

In many cases the delivery of services to citizens and firms requires synchronization of government activities at the national level (e.g. register of enterprises and social security), the regional level (e.g. environment, education and healthcare) and the local level (e.g. building permits and some taxes). They are all considering establishing One Stop Shop concepts, such as for enterprises. However, in the 1990s cooperation between government layers in such fields has been hampered by conflicting political affiliations at city (socialists), regional (nationalists) and national levels (liberals and conservatives).

International Context

Barcelona is a city with a high international profile. With the organization of the Olympic Games in 1992 it put itself on the world map. Since then, the city government has maintained an international perspective. Mayor Pasqual Maragall was one of the founders of the Eurocities network, an interest group of the major European cities. A few years later Barcelona – under the leadership of vice-mayor Ernest Maragall – was amongst the founding members of the Telecities network for ICT and city development.

Many parts of the municipal website are not only translated into Castellan, but also into English.

The international profile of Barcelona attracts many foreigners to the city. Many people visit the city as tourists or for business. Many foreigners have also moved to Barcelona to live. Table 3.3 reveals that more than half of the foreigners who live in Barcelona have come from the American continent. Most of them (98 per cent) are from Latin America. The number of immigrants to Barcelona has risen by more than 40 per cent over the last year.

4 Barcelona's Vision on E-Governance

The Municipal Organization

The city of Barcelona complies with the third strategic plan for Barcelona Metropolitan Region and the action lines it contains. The Municipal Action Programme 2000–2003 elaborates the municipal dimensions and highlights 'the political will to make Barcelona City Council an example of good government and good administration'.[1] The action programme identifies six priority lines of action:

1 City Council online;
2 attention to citizens and quality of service;
3 receptivity, participation and consensus;
4 quality of public areas;
5 organizational development;
6 new management tools.

---

[1] *Programa d'Actuació Municipal 2000–2003*, City of Barcelona (2000).

Table 3.3 Number of foreigners in Barcelona, 1996–2003

| | Number of foreigners in Barcelona, 1996–2003 (beginning of the year) | | | | | | | | |
	1996 N	1999 N	2000 N	2001 N	2002 N	2003 N	2003 %	2002–2003 increase
European nationalities	9,407	11,289	13,316	16,286	22,924	33,111	20.3	44.4
Asian nationalities	5,576	7,164	9,326	12,175	17,934	26,412	16.2	47.3
African nationalities	4,044	6,365	8,181	10,044	13,983	17,002	10.4	21.6
American nationalities	10,213	15,972	22,488	35,378	58,944	86,364	53.0	46.5
Oceanic nationalities	76	81	85	99	114	157	0.1	37.7
Total number	29,316	40,871	53,396	73,982	113,899	163,046	100.0	43.1
As a % of total population	1.9	2.7	3.5	4.9	7.6	10.7		

Source: Barcelona Statistical Department (2003).

These action lines have been further elaborated in the Municipal Innovation Plan.[2] The Innovation Plan constitutes a common philosophy and a strategy for attaining a dynamic municipal organization that offers innovative solutions to citizens' needs, as required by the City Council. It aims at maintaining the municipal organization in a leading international position on issues relating to administrative modernization, quality of service and accountability to citizens.

The Innovation Plan also aims to be an instrument that is both supportive and demanding of managers with regard to the improvement of services and sending a message to employees that stimulates their participation. It strives for a corporate identity and pride amongst the members of the municipal organization.

The city council and administration of Barcelona employ more than 12,000 people. The organization has a well decentralized structure. There is a territorial decentralization into 10 city districts and a functional decentralization into five main departments. These central departments focus on town planning (including public works), urban services (including utilities and environmental services), mobility and safety (including urban police, fire brigade, traffic control), social services (including welfare, education and culture) and financial relations (including tax collections). A number of more or less autonomous institutions and enterprises are linked to the municipal departments. The municipal departments have 6,753 employees and the municipal agencies and companies 5,291.[3]

In addition the General Services Department comprises such services as organization, personnel, accounting, municipal buildings, customer relations and ICT.

The five main departments are replicated in each of the 10 city districts. Not only day-to-day decisions, but also the strategic decisions are taken in the districts. The role of the central departments is mainly in process design, standardization of services throughout the municipality, quality control and overall productivity.

Integrated attention to the citizens is one of the spearheads of the Barcelona approach to the inhabitants of the city. All relations between the city and the citizens are taken care of on an exclusive basis by a 'one stop shop' service. This service, Barcelona Informacio (BI), is a relatively autonomous institute that reports to the director of General Services. It is strongly supported by the executive city management and the political leaders of the city government. By consequence there is no direct contact between the main city departments and the citizens – with the exception of specialized services.

The Municipal Innovation Plan

In 1998 the IT strategy for Barcelona 1999–2003 was established. This strategy had four main components:

---

[2] *Municipal Innovation Plan*, City of Barcelona (2002).
[3] *Annual Report* 2000, City of Barcelona.

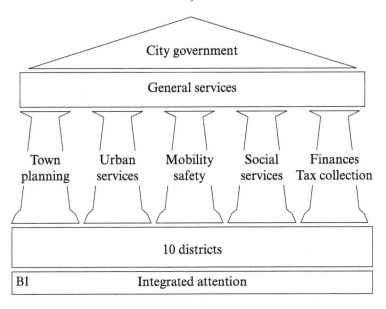

Figure 3.3   Municipal structure

Source: Olivella (2000).

1   an integrated information platform. All types of information are managed in a coherent way;
2   a reliable technological infrastructure (computer systems and networks). The infrastructure is needed to deliver good quality services to city employees and external customers (citizens and enterprises);
3   internal process re-engineering. Business processes benefit from facilities such as intranet, electronic workflows and digital signatures;
4   online services delivery. This is considered as the most important component, since it brings direct benefits to the citizens and the enterprises in the city.

The Municipal Innovation Plan 2002 regrouped these components into the objectives listed overleaf. These objectives and action lines can be found under different strategic action lines of the Municipal Information Plan, which also identify a number of already-running projects.[4]

The Barcelona website has a direct link to the overall city policies. Mayor Joan Clos plays an important role. Every Monday he evaluates the website with Barcelona Informacio. Initially the mayor focused on improving the sites basic presentation and services. The focus has gradually shifted to include creating an image for the city.

---

[4]   A full list of projects under these objectives at the time of writing is in Annex 1.

Action Line: City Council Online
- Promote Internet culture amongst the population and within the City Council (1 project)
- Increase the range of online public services (20 projects)
- Develop individualized contents (9 projects, including some for small and medium sized enterprises)
- Improve the security of online procedures, promoting the adaptation of the legislative framework (1 project)
- Facilitate the collaboration with other administrations, suppliers, etc. through the Extranet (2 projects)
- Facilitate the input of information to online services databases to improve attention to citizens (2 projects)

Action Line: Receptivity, Participation and Consensus
- Create new formulas for citizens participation (12 projects, including online participation and electronic voting)

Action Line: Organizational Development
- Share the knowledge of professionals and foster a transversal approach to improve internal management and attention to the citizens (13 projects)

Action Line: New Management Tools
- Reduce paperwork, via online internal management (16 projects)
- Exchange of information between sectors with shared applications and databases (6 projects)

Although Mayor Clos and Councillor Maragall – and the directors of IMI and BI on his behalf – have the power to negotiate, they follow a consensus policy to improve electronic government and service delivery. It is a joint project of all involved. IMI, BI and the department responsible for a particular service work together to make the website work.

Barcelona is big enough to follow its own course (which it does), but at the same time it tries to synchronize activities at the regional level. *Localret* is an exponent of these efforts. It is an organization that unites the municipalities of Catalonia and supports the development of infrastructure, content and access to e-government in Catalonia.

The Localret consortium currently includes 751 municipalities, which cover 98.3 per cent of the Catalonian population, the four provinces of Catalonia, and both of Catalonia's municipal associations: the Federation of Catalan Municipalities (FMC) and the Catalan Association of Municipalities and Counties (ACMC). It is supported by the regional government. The city of Barcelona plays a crucial role in Localret. Councillor Ernest Maragall took the initiative and is the current vice-president.

E-Governance in Barcelona

Information and communication technologies play an important role in realizing the Municipal Innovation Plan and in maintaining the integrated attention concept. The city government decided that integrated attention to the citizens and territorial and functional decentralization require a centralized ICT strategy and centrally managed facilities.

The municipal institute for informatics (IMI) is an autonomous institute, 100 per cent owned by the municipality, that is attached to the director of General Services. IMI employs 195 persons. Figure 3.4 illustrates the distribution of these employees over the different departments. The annual turnover is €21m, which are spent on a not-for-profit basis on personnel (€10m), investments (€6m) and other expenses (€4.8m).

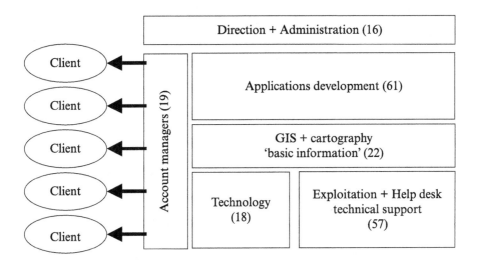

Figure 3.4   Internal structure of IMI

Source: Olivella (2002).

Although municipal departments are free to choose other IT service providers, IMI has a de facto monopoly. There are two main reasons for this. First, much like Barcelona Informacio, IMI is strongly supported by the executive city management and the political leadership of the city. Second, in order to allow functional and territorial decentralization, as well as integrated attention to the citizens, almost every IT system and application must fit in the highly integrated infrastructure of equipment, software and information.

Every year IMI has to negotiate contracts with each of the municipal departments. These are complex negotiations, because the partners have to find a balance between the clients' priorities as represented by the department managers and the global interests of the city as represented by the director of the General Services Department. The latter adheres to the strategic priorities of the mayor and the city council.

Recently IMI has decided to put more emphasis on innovative services for Barcelona, concentrating on Internet and intranet developments. It has begun to outsource activities that are available in the market at a competitive price and quality. The information and systems strategy, however, will always remain in IMI, at the core of the municipality.

This is quite different from the approach of the regional government of Catalonia. This 'Generalitat' outsourced its entire IT department to a subsidiary of T-systems, a Germany-based IT company. Only a small ICT policy unit remained at the government level. But the situation of the outsourced IT department was a little chaotic at the start, because many experts left the organization. This has hampered the development of a regional ICT strategy, which is a process that requires continuity and endurance. Consequently, cooperation between regional and local government has been difficult. Since the mid-1990s, however, the Catalonian parliament has been pushing for a synchronized local and regional ICT strategy, with high ambitions. *Localret* is an exponent of this era.

5 Governing Content

Over the years the Municipal Institute for Informatics (IMI) has built comprehensive information architecture. At present a geographic information system (GIS) is central to this architecture, as can be seen from Figure 3.5.

Barcelona's Internet strategy serves a number of simultaneous goals:

1   improve services to the citizens, including the use of the municipal website www.bcn.es;
2   improve efficiency in city hall, including consensus between the different municipal departments, and the use of internet and intranet by municipal employees;
3   improve citizens' participation;
4   improve image of the city, including the integration of tourism in municipal attention.

Online services include the provision of information about services and legislation, online transactions, access to libraries and the diffusion of information via e-mail. They are enabled by a system of content management, an electronic city agenda, a digital city map (GIS) and other elements of the information architecture.

The Municipality Online initiative is not only used to improve services towards citizens, but also to reorganize municipal organization. Municipality Online was introduced in several phases. In the first phase only information was presented

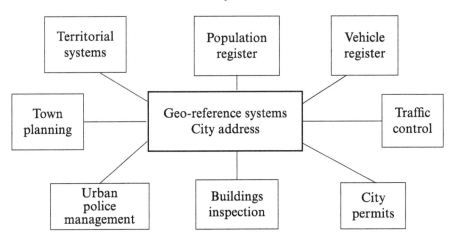

Figure 3.5 Integrated information architecture

via the Internet. As a next step some services were put online. The third phase envisioned the integration of Internet services with the municipal intranet. Finally, all administrative levels were connected.

Electronic participation and e-democracy are the least developed parts of the municipal Internet. There is a discussion platform on www.bcn.es, but in the beginning of 2002 there were not many contributions out there. As to e-voting, a pilot project for specific and concrete issues was launched in 2002. The expectations are not very high. Some people use it, some do not. The local government of Barcelona fears that this may result in biased outcomes of the e-voting process.

This raises the issue of becoming familiar with using the Internet. One way to stimulate citizens to get used to the Internet is to stimulate them to put their own content on the web. This would, for instance, support neighbourhood initiatives and other grassroots initiatives to participate in the information society. Although some initiatives of this kind occur in Barcelona, the idea of neighbourhood websites is too far away yet from the formal decision process in the municipality. Already existing community websites are not linked to the municipal website.

One of the most eye-catching phenomena in the way Barcelona organizes its content to the citizens is the integrated approach towards telephony, Internet and district offices. Barcelona Informacio uses three main channels for citizens relations:

- a call centre, created in 1986, which gives access to all services through the local telephone number 010. In 2001 the call centre handled an average of 16,000 incoming calls per day;
- integrated offices in the city districts, created in 1990. In 2001 an average of 10,000 people per day visited these offices;

- Internet services at www.bcn.es, created since 1995. In 2001 an average of 37,000 people per day visited the municipal website.

Councillor Ernesto Maragall was the leading initiator for the call centre and a number of other projects.

The 010 Call Centre

The 010 call centre is ISO standardized. It receives many telephone calls, 80 per cent of which are handled by an external agency. The quality of the call centre is evaluated on a continuous base, using surveys and several other assessment techniques. The 010 call centre receives about 16,000 phone calls each day. About 16 per cent of the calls ask for municipal information while many of the remaining phone calls have to do with leisure-related questions.

Barcelona Website(s)

In the beginning the Internet services were targeted at tourists. It was a joint project of the municipality and the technical university. From 1997 onwards the website has acted as an information channel to the citizens. Interactive tools for service transactions and citizens participation were introduced in 1999. Since 2000 the website has covered all aspects of municipal content and services.

Figure 3.6 shows the development in the use of the municipal Internet and call centre over the last five years. It shows that the introduction of interactivity in 1999 has resulted in an accelerated growth of the use of municipal Internet services. However, this did not lead to a reduction in the number of call centre contacts.

The most popular parts of the website are the city guide, city directory and agenda (35,000 visits per week), urban planning maps (15,000 visits per week), transport and traffic information (12,400 visits per week). Since 2000 customer surveys have been held on a regular basis. These surveys reveal that citizens find the content of the municipal website to be interesting. Citizens found the number of services available on the website to be sufficient. The navigation and the design of the website were rated as weak points. Different departments use different website layouts. The citizens ask for a comprehensive website. On the basis of these findings efforts have been put in place to harmonize the layout and content of the website. A content management system is used (*Vignet*) and www.bcn.es acts as a link to other websites that use the same layout.

At present the municipal website www.bcn.es offers 25,000 pages of static information and contains the services of 60 departments in the city of Barcelona. Tourist information is also included. Furthermore, Barcelona's website already provides some information about activities in other Catalan municipalities. However, there is not much about healthcare (such as pharmacies), going out (e.g. cinemas), enterprises and education. These services are managed by other

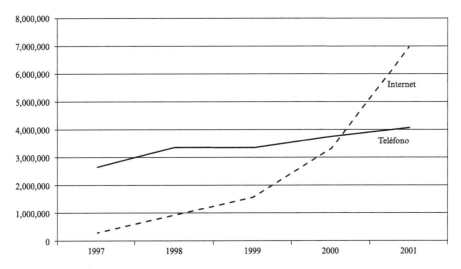

Figure 3.6 Use of the municipal call centre and website, 1997–2001

Source: IMI, 2002.

institutions and government levels. Integrated services require joint inter-organizational efforts.

Several project are underway to realize *Internet portals* that cross the organizational boundaries, putting the customer in focus. *Localret* is an important initiative in this field. At the regional level *Localret* is involved in several activities, such as the 'Open Administration of Catalonia' project (AOC), which should lead to a single e-government portal for all municipalities in the Catalan region. This project includes the introduction of digital signatures. After testing feasibility amongst 40,000 citizens in the municipality of Barcelona and the Catalan region, the project – if proven successful – will be widened to a full scale of implementation.

Meanwhile the municipal website of Barcelona offers an online *citizens' dossier*, a personalized portal that makes personal data from different municipal systems accessible to the individual. The personalized website uses a digital signature as an authorization to access the data. The system is already available. However, only a small number of inhabitants have a digital signature yet. The municipality does not like to impose an aggressive strategy to promote the use of digital signatures by the citizens. Alternative ways to get the services delivered – not making use of the personalized portal and a digital signature – are being kept open at all times.

The integration of services that need input from different government levels is often hampered by an insufficient level of standardization and data harmonization. In the Spanish case, for instance, the national tax office and the national social security register are using incompatible digital signatures, which is not very helpful for effectuating socio-economic strategies at local level.

6   Governing Access

Creating Internet access opportunities was part of the strategic plan for the information society in Barcelona. However, key actors in the field of *public access and training* belong to different administrative entities. The libraries are operated by the province, the schools fall under the region. In addition some libraries and schools in Barcelona belong to the municipality. In the second half of the 1990s the different actors did not succeed in developing joint actions for public access. By consequence, public access was left to the market. Apparently, the market did well. There are remarkably many Internet cafés in Barcelona, many of them owned and operated by *Telefonica*.

In January 2001 the municipality (Ajuntament de Barcelona) and the province (Diputacio de Barcelona) created a consortium to enhance Internet access in libraries.[5] The consortium develops a user-friendly Internet portal for library services. Furthermore it wants to provide all the users of Barcelona's libraries with public access to the Internet. This involves installing computers in the libraries. Internet training courses have been introduced in 2002 as a way of attempting to reach all target groups of libraries.

A survey[6] shows that by the end of 2001 39.7 per cent of the Barcelona inhabitants of 16 years and older used the Internet. Table 3.4 is taken from the survey and shows that 10 per cent of Internet usage is from 'other places', which would include Internet cafés and libraries.

There are efforts to increase Internet access in municipal schools. Each municipal school has a website. Furthermore, people from the neighbourhood can reserve computers after school hours for a range of activities.

The Department of Youth teaches children how to use computers and the Department of Elderly People gives courses to elderly people. The youth courses are the most advanced.

7   Governing Infrastructure

Under Spanish law each regional area was given the opportunity to establish the conditions for a second cable telecommunications operator.[7] However, it did not work out well in Catalonia. Although the second operator, Cable Television de

---

5   Consorci de Biblioteques de Barcelona.
6   Published at http://www.bcn.es/english/estadist/itaules.htm.
7   Spanish Law 42/1995 divided Catalonia in three areas for the provision of cable telecommunications services: the Northeast Area, the Western Area and the Barcelona-Besòs Area.

Table 3.4 Location of Internet use

	Places where people use the Internet (%)					
	June 1999	December 1999	June 2000	December 2000	June 2001	December 2001
Home	51.3	56.5	66.4	65.2	77.7	70.0
Work	40.9	45.3	46.6	43.0	42.6	40.1
School/ place of study	19.7	21.1	20.5	16.7	14.7	14.1
Others	4.7	4.9	2.6	8.2	3.6	9.8
Don't know/ no response	0.0	0.0	0.0	0.3	0.0	0.3

Source: Barcelona Statistical Department.

Catalonia (CTC), was supported by Barcelona municipality, it still has several problems. One of the main reasons is that during the preparations for the 1992 Olympic Games Telefonica invested heavily in fibre. This turned out to be a strong competitive advantage when telecommunications were liberalized in the mid-1990s. Furthermore, when cable television was introduced Telefonica decided to distribute the national television channels via satellite, thus taking the wind out of the cable operator's sails.[8]

In 1997 Catalonian local authorities (cities and provinces) established a regional organization called Localret. This consortium was supposed to guide the tendering process for new operators, with the ambition to increase economic and social development at both the local and the regional level, whilst at the same time coping with potential threats to social cohesion and regional balance.

The scope of the consortium's activities has steadily widened. On the one hand the dominant position of Telefonica hampered the market development for second (cable) operations, thus demanding other ways to address the issues at stake. On the other hand, the broad bearing of Localret made it an ideal vehicle for fostering information society issues and public initiatives involving local government and telecom operators. Thus, the main issues to address were:

- On what terms would new telecommunication networks be built and used in Catalonia?
- What role should information technology play in enhancing public services and citizens' quality of life?

---

[8] Note that *La Caixa* (a Spanish bank with a strong position in Catalonia) and the regional bank of the Basque region each own 5 per cent of the shares of Telefonica.

Several infrastructure operators are active in the Catalan region at the time of writing. On behalf of its members Localret tries to manage *right of way* issues, such as the time schedules for operators to lay cables in the ground. In one such effort Localret has put empty tubes in the streets of Barcelona. Five out of the six operators at the time took the opportunity to put their cables in these tubes at the same time.

Although infrastructure is mainly an issue at regional level, Barcelona deals with most of its infrastructure itself. There are about 250 km of empty tubes for fibre connections underground in Barcelona. Some 80 municipal buildings have already been connected to a broadband fibre-optic infrastructure, which is owned by the city and operated by IMI.

8   Conclusions and Recommendations

It is clear that the electronic government strategy of Barcelona is extremely well organized. The information systems architecture has been developed over the years. It is a solid architecture. Furthermore, although legacy systems are no doubt present in the administration, the Municipal Institute for Informatics (IMI) appears to be successful in matching those with the requirements of today: interactive and integrated applications.

The IT strategy of Barcelona supports an evenly well-developed business strategy towards the citizens. Integrated attention on citizens is a strong concept and the department of Barcelona Informacio (BI) makes it work remarkably well in Barcelona.

The leading professionals in both departments – IMI and BI – appear to be on the same wavelength. They work together quite well. This is a key factor for the success of the Barcelona electronic government practice so far. It will prove to be of great value for the next steps in the development of transparent and open government in Barcelona. These steps may include a higher level of use of interactive services, electronic participation and neighbourhood-based content on the web.

However, the most decisive factor in the success of the Barcelona case has been the consistent support and commitment from key politicians. Councillor Ernest Maragall has been a constant factor all the way from the opening of the call centre in 1986 to the present day. His vision and guidance have kept equal pace with the development of new technologies over the 1990s. He made it happen, and the professionals from BI and IMI have always been there to make it work. The personal attention of Mayor Joan Clos, who evaluated the municipal website every Monday morning, has also been a key factor to motivate those who were involved.

The development of the municipal call centre and the municipal websites in Barcelona appear not to have been driven by customer demand. This is not unusual for new technologies. There is always a demand for service and service quality, whilst there is seldom a demand for specific instruments or technologies to deliver the services. In other words: there is a lack of demand for IT services, even if the technological capacity is available. However, the statistics presented in this report reveal that the use of IT services in Barcelona has increased dramatically after their initial introduction, thus revealing a considerable potential demand.

A key factor in unleashing potential demand is effectiveness and efficiency in sectors like education, health and social services. These services are highly interactive, but in many cases the administrative structures are not very flexible, and the employees hardly use interactive technologies. At the same time, the demand for these services tends to increase. Changes in administrative procedures, empowerment of civil servants and effective use of technology (including mobile and broadband facilities) would enable the city to cope with the challenge.

There is room for more use of Internet and intranet in the municipality. Young municipal employees tend to be used to those instruments, but older people seem to have a more traditional approach to their work. When intranet applications are being introduced in the administration, particular training efforts should be developed for these groups of senior employees. This is in line with the Human Resource approach to management: 'investing in people, improving employability'.

Public and Private Partnerships

In many cases the integrated services that information and communication technologies can deliver – if they were to be exploited to the full – require effective interaction between public and private organizations. We have not found much evidence of such forms of cooperation in Barcelona:

- At the level of *content* the municipal website reflects the municipal administrative structure. Apparently, it does not follow so-called lifetime events (e.g. birth, going to school, looking for a job, marriage, death) and it does not relate to private partners who could provide services for these events.
- At the level of *infrastructure* there is no evidence of effective partnerships with private operators that exceed the level of customer-supplier relations. This may well be the result of the Spanish way to allow a 'second operator' at regional level. The governments of Barcelona and Catalonia could not resist the temptation to start their own regional cable operator – rather than opening up the market to foreign competition.
- At the level of *access* the market (including *Telefonica*) has taken the initiative to create Internet cafés throughout Barcelona. The municipality and the province of Barcelona started with interesting initiatives in libraries and schools. Although there is still room to improve the level of computer education and the number of public access points, we have heard of no private sector efforts to support these local public initiatives.

In any case, there is a potential to be exploited which could contribute to the promotion of an Internet culture amongst the population and within the City Council. This is the first action line mentioned in the Municipal Innovation Plan.

The Municipal Institute of Informatics (IMI)

The Municipal Institute of Informatics (IMI) is an impressive organization. It spends 50 per cent of the IT budget on technological infrastructures. This seems reasonable, since content can only be organized and accessed in an optimal way if a solid infrastructure exists.

This organization produces a large output with relatively few resources available. IMI manages 252 applications. The level of quality is good and the organization is quite efficient. It remains to be seen, however, what will happen when additional resources have to be allocated to information and communication technologies (ICT). This need may arise as a result of the Municipal Innovation Plan. Even in the case of Barcelona – with a high level of political interest in services and good government – technology is often considered to be 'just a tool'. The technological infrastructure is invisibly hidden behind the scenes and organizational workflows.

Recently, Barcelona has begun to consider outsourcing the operational activities of IMI. The information and systems strategy, however, would remain in IMI, at the core of the municipality. We believe that this issue should be approached with the utmost caution. The Municipal Institute of Informatics ranks among the best of its kind in Europe. This is only partly based on the vision and the strategy towards citizens, municipal organization and the use of technology. There are many more cities in Europe with a similarly excellent strategy. However, they are less successful than Barcelona in implementing the strategy, because they lack the history of technological experience as well as the direct access to – and control over – the application developers, the operators and the engineers who *make it work*. Many of those who outsourced their IT department in the past have regretted it. The city of Leipzig even bought back its IT department a few years after it had been outsourced and modelled its new position in the municipal structure to the Barcelona example. One of the issues at stake is that sooner or later the mother company of the outsourced IT department seeks to impose its own standards of organization, technology and information architecture. Furthermore, in the free market there is always a risk that ownership of the outsourced IT department or its mother company will change.

Moreover, one of the main threats to the integrated information architecture – in Barcelona too – lies in the advice that business consultants tend to give to their customers. It is not unknown for them to give an unbalanced view of the corporate municipal interests, the business unit's perspective and their own scope of experience. They could also listen too closely to the mother company of the outsourced IT department. In many cases municipal ownership and excellent, up-to-date knowledge of the technological implications is a last remedy to cope with these kinds of threats to the concept of integrated attention to the citizens, the technological, organizational and political requirements.

National and Regional Perspective

Cooperation between government levels is important. This is even more true for the metropolitan regions. Due to their scale and their socio-economic fabric they have to cope with issues that could formally belong to the jurisdiction of any administrative level – be it municipal, provincial, regional or national. Integrated attention to citizens and integration of services on technical platforms – such as the Internet – put much stress on the issue of intergovernmental cooperation. Both the integration of services and intergovernmental cooperation require a high level of standardization of services and technologies used. It would be helpful if the national, regional and local levels would engage in an effort of harmonization. The challenge here is to do this on a basis of equality: national and regional levels should have an open ear and an open mind to the experience and needs at local level, whilst the local level should be willing to exchange some of its specificity for a better integration from the customer's point of view. Stimulating networks of exchange between government officials could be an effective way to achieve this.

References

Barcelona City Council (2000a), *Barcelona Digital City.*
Barcelona City Council (2000b), *Programa d'Actuació Municipal 2000–2003.*
Barcelona City Council (2001), *Annual Report 2000.*
Localret (2001), 'Informe sobre el desplegament de la xarxa de banda ampla interactiva a Catalunya', available at www.localret.es/societat/docs/ica15600.pdf.
Olivella, Lluís (2000), *Tecnologies de la informació i modernització de la ciutat*, Aula Barcelona, Barcelona.

Web Pages

Barcelona City Council: http://www.bcn.es.
Barcelona Activa: http://www.barcelonactiva.es.
Catalan Institute of Statistics (Idescat): http://www.idescat.es.
Generalitat de Catalunya (Government of Catalonia): http://www.gencat.net.
Localret: http://www.localret.es.
Project 22@ bcn: http://www.bcn.es/22@bcn.
Statistic Department, Barcelona City Council: http://www.bcn.es/estadistica.

Interview partners

Joan Battle, IMI, Coordinator, EU projects.
Marta Continente, Barcelona Informacio, General Manager.
Lola Gonzalez, General Services Department, Director.
Ernest Maragall, Councillor.
Lluis Olivella, Municipal Institute for Informatics, General Manager.
Jordi Pericas, Localret, General Director.
Isabel Ricart, IMi, Account Manager.
Llouis Sanz, IMI, Assistant Director, Information and Cartography.
Teresa Serra, Member of the Parliament of Catalonia, President of the Information Society Committee.

Annex 1 Selection of IT projects in the Municipal Innovation programme

1   City Council Online

1.1   Promote Internet culture amongst the population and within the City Council

Cibernarium   Internet multi-space dedicated to digital learning through flexible training and experimentation formulas.

1.2   Increase online public services

Getting Around Barcelona	Interactive website for facilitating travel on public transport, in private vehicles, on foot or by bicycle.
New structure for the website	New general layout of the website to improve its presentation, navigability, user-friendliness and conceptualization of the web contents.
Procedures online	Over 60 procedures that can be carried out online.
Meeting venues online	Centralised information and reservations via the municipal website, for hiring municipal venues.
Municipal Council meetings from home	Direct and delayed access to the plenary sessions of the Municipal Council via BTV and the website.
Regulations online	All the municipal regulations, regularly updated and easy to access.
Municipal Digital Images Bank (BDM)	Bank of images of the city accessible via the web.
Notifications online	Notification of fines for vehicle hire companies using internet channels.
Urban planning projects in view	Public information on urban planning projects on the web.
Virtual access to specialised libraries	Integration of all of the specialist municipal library catalogues (Cultural Institute – ICUB, Municipal Education Institute – IMEB, City Hall).

1.2   Increase online public services (cont'd)

Archive online	Integrated access to the 14 Municipal Archive centres, with access to publications, consultation of documents and databases, and the possibility of sending requests.
Internet Call Centre	A telephone helpline for resolving difficulties in using the website.
E-mails: immediate attention	Municipal call centre for the automatic management of mail sent to the municipal website mailboxes.
Complete notary procedures online	Online processing of all the municipal procedures required following the transmission of goods.
Tax appeals online	Presentation of tax appeals and applications through internet channels.
Municipal schools website	Website of all the municipal education centres in Barcelona, to inform upon all the services and educational activities at each school.
Online matriculation process	Consultation via the Internet regarding the matriculation process for municipal education centres.
Education services online	Online service for the education centres in the city and its area of influence to allow consultation and reservation of educational services.
Online school libraries catalogue	Integrated catalogue of municipal primary school libraries, accessible via the Internet.
Online communication with Neighbourhood Associations	Internet-based communication between the District and its neighbourhood associations.

1.3   Develop personalised contents

Barcelonanetactiva (BarcelonaActiveNet)	Portal of services and contents for new entrepreneurs.
Barcelona Business	Areas with services and contents for attracting foreign investment.
Infopime (InfoSME)	Internet service containing information, relations and procedures addressed to the city's SME sector.

1.3   Develop personalised contents (cont'd)

Virtual restoration office	Website that offers all the information necessary for restoring an older building.
DSI: internet	Selective dissemination of information (DSI) via e-mail to citizens, organizations and companies that express specific interests.
Culture Channel	Cultural multi-channel in suitable format for the Internet, mobile telephones and television.
Barcelona public art online	Online catalogue of Barcelona's public art.
Taxpayers' website	An extranet that offers taxpayers integrated information on their tax situation and allows virtual procedures to be processed.
Digitalization of specialised libraries	Digitalization of specialised libraries with collections of local interest or for the specialist public, in order to facilitate access to them via the web.

1.4   Improve the security of online procedures, promoting the adaptation of the legislative framework (electronic signatures, etc.).

Electronic administrative register	Applications, petitions and allegations via the Internet.

1.5   Facilitate collaboration with other administrations, suppliers, etc. via the Internet

PIRMI online	Application for the inter-administrative processing of the Minimum Salary Insertion Plan.
Suppliers' homepage	Extranet offering companies that work with the City Council information on tenders and invoicing status.

1.6 **Facilitate the supply of information to improve attention to citizens**

New contents manager
for the website
À.S.I.A.

New platform for creating portals intended to allow a more efficient and secure management of municipal websites.
New databases for amenities, agenda and further procedures.

3 **Receptivity, participation and consensus**

3.3 **Create new formulas for citizen participation, adapted to current changes and taking into account new collectives and social agents**

Virtual participation
Barcelona under debate

Citizen consultations, chats and forums via the Internet.
Permanent online space for debate regarding issues of interest to citizens and accessible to any interested party.

Eurocity: electronic voting

First European experiment of smart cards that will allow greater insight into electronic democracy experiences.

5 **Organizational Development**

5.4 **Share the knowledge of professionals and foster a transversal approach to improve internal management and attention to citizens**

The Information Providers
Portal

On the municipal intranet homepage incorporate direct access to applications for both City Council information providers and other municipal workers carrying out their normal functions.

Info Bib

Electronic bulletin for circulating information of interest to the City Council's managers and professional staff.

Knowledge within reach

The Municipal Library as a made-to-measure information research centre.

5.4 Share the knowledge of professionals and foster a transversal approach to improve internal management and attention to citizens (cont'd)

Integrated opinion	Improvement and interconnection of the City Council's public opinion radars (Department of Studies, Complaints and Suggestions, Barcelona Information, Participation Councils, Councillors, media analysis, etc.).
DSI-Intranet	Selective dissemination of information to municipal workers in accordance with their profiles and interests.
Virtual resources for innovation	Questionnaires and instruments of self-assessment for managers, selection of texts and webs on innovation in the public sector and mailbox for collecting suggestions for innovation from employees.
Intranet	New home page, new design and new utilities on the corporate intranet, configured as a single portal for all the workers.
Paper-free personnel procedures	Processing of personnel administration via the intranet.
Virtual communities	Virtual communities for sharing knowledge and good practice amongst professional collectives (lawyers, inspectors, personnel services, etc.).
BCN educational boulevard	A virtual communication space for the network of municipal schools.
Internal health guidance plan	Internal plan for health guidance addressed to Urban Police personnel.
Innova	Information application for the management of the City Council's database of innovative projects.
G.U. (Urban Police Force)	Circulation via the intranet of circulars, orders, etc., aiming towards paper-free management and achieving accessibility.

6.1 Reduce paperwork via online internal management

GPS local police	Installation of a satellite monitoring system for a more efficient assignation of services for the local police vehicles.
GPS tow-trucks	Installation of a system of satellite monitoring to improve the system of assigning services to tow trucks.
E-pocket	Integral control of public areas (CIEP) via a detection system with an e-pocket camera, which allows mechanised data capture.
Photocontrol	Photographic control of traffic light infringements through the use of digital cameras.
Automatic speed traps	Automatic speed traps with digital photograph on the ring roads.
Electronic fines	Electronic register of traffic infringements in real time via pocket computers.
Controlled litter bins	Control of the state of litter bins using a chip.
Digitalised invoices	Digitalization of invoices to speed up payment orders.
Electronic invoices	Invoices for consumption of basic services by computer.
Copèrnic	Documentation for the Executive Committee meetings and those of other government bodies available from the intranet.
Paper-free travel	Processing of travel documents using electronic signature.
Bar code on notifications	Inclusion of labels with bar codes on notifications, to stick to related appeals and documents presented by citizens.
Identified premises	Creation of a single identifier for business premises, that is used for feeding into all fiscal databases.
Digitalized urban planning documentation	Digitalization of the graphics collection and planning documentation in the Urban Planning sector.

6.1 Reduce paperwork via online internal management (cont'd)

Queue managers
Control of waiting time with queue managers at all offices that attend citizens and experience mass affluence.

Office of Information and Procedures of the Urban Police Force
Creation of an integrated office of information and procedures for the Urban Police Force.

6.2 Exchange of information between sectors, with shared applications and databases

Virtual Library
Access for municipal workers to any type of information resources through a single documental search in the General Library catalogue.

AIDA
Implementation of a computerised documentation management system at the Municipal Archive.

Municipal Amenities Master Plan 2010
Production of integrated maps and establishment of criteria for the progressive rationalization and compacting of amenities.

Weekly Agenda
Weekly agenda of events affecting public roads for the programming and coordination of municipal services.

New permits accounts model
Link-up between the processing database and the accounting management database for permits.

Let's Remember Barcelona
Digitalization of the historical documents of the Municipal Archive, with private sponsorship.

Annex 2 Data on computer and Internet use in Barcelona, 1997–2001

At home ... (Summary) (%)

	September 1997	March 1998	September 1998	June 1999	December 1999	June 2000	December 2000	June 2001	December 2001
Have Internet at home	6.4	10.8	11.3	14.7	18.9	26.8	32.2	37.3	38.3
Have computer but not the Internet	37.0	34.2	35.0	35.7	32.8	27.2	25.8	20.6	20.0
Do not have computer	56.6	55.0	53.7	49.6	48.3	46.0	42.0	42.1	41.7
No.	(1,000)	(1,000)	(1,000)	(1,000)	(1,000)	(1,000)	(1,000)	(1,000)	(1,000)

1 Do you have computer at home? (%)

	September 1997	March 1998	September 1998	June 1999	December 1999	June 2000	December 2000	June 2001	December 2001
Yes	43.4	45.0	46.3	50.4	51.7	54.0	58.0	57.9	58.2
No	56.6	55.0	53.7	49.6	48.3	46.0	42.0	42.1	41.7
Don't know/no response									0.1
No.	(1,000)	(1,000)	(1,000)	(1,000)	(1,000)	(1,000)	(1,000)	(1,000)	(1,000)

2 *And do you intend to get a computer in the next 12 months? (Only for those who do not have a computer) (%)*

	June 1999	December 1999	June 2000	December 2000	June 2001	December 2001
Yes	10.3	13.9	11.1	13.3	10.5	11.8
No	88.3	84.5	84.6	85.5	86.9	84.9
Don't know/no response	1.4	1.7	4.3	1.2	2.6	3.4
No.	**(496)**	**(483)**	**(460)**	**(420)**	**(421)**	**(417)**

3 *How many computers do you have at home? (Only for those who have a computer at home) (%)*

	June 2001	December 2001
One	77.9	78.2
Two	18.7	16.7
Three	2.6	3.4
Four	0.9	1.4
Five		0.2
Six		0.2
No.	**(579)**	**(582)**

4 *Is the computer in your home a desktop or a laptop? (Only for those who have a computer at home) (Multiple choice)**

	September 1997	March 1998	September 1998	June 1999	December 1999	June 2000	December 2000	June 2001	December 2001
Desktop							88.6	89.1	87.8
Laptop							4.3	10.9	12.0
Both							6.2	–	–
Don't know							0.9	0.0	0.1
No response							0.0	0.0	0.0
No.							(580)	(579)	(582)

* In December 2000 this question was not multiple choice.

5 *Can you tell me what model it is? (Only for those who have a computer at home)*

	September 1997	March 1998	September 1998	June 1999	December 1999	June 2000	December 2000	June 2001	December 2001
Pentium	30.2	42.4		54.4	59.0	62.6	60.7		
486	19.4	15.6		7.3	5.8	6.1	7.1		
PC								83.8	85.5
Macintosh/Apple	4.8	2.7		2.4	3.5	3.3	2.6	3.7	2.9
Other	16.4	2.9		3.8	1.9	5.7	3.1	0.7	0.3
Don't know/no response	29.3	36.4		32.1	29.8	22.2	26.6	11.8	11.3
No.	(434)	(450)		(507)	(517)	(540)	(580)	(579)	(582)

6 *And at home are you connected to the Internet?*

	September 1997	March 1998	September 1998	June 1999	December 1999	June 2000	December 2000	June 2001	December 2001
Yes	6.4	10.8	11.3	14.7	18.9	26.8	32.2	37.3	38.3
No	93.6	89.2	88.2	84.9	80.3	72.9	66.1	62.2	60.6
Don't know/no response	0.0	0.0	0.5	0.4	0.8	0.3	1.7	0.5	1.1
No.	(1,000)	(1,000)	(1,000)	(1,000)	(1,000)	(1,000)	(1,000)	(1,000)	(1,000)

7 *Do you intend to get connected to the Internet in the next 12 months? (Only for those who do not have Internet at home)*

	June 1999	December 1999	June 2000	December 2000	June 2001	December 2001
Yes	19.8	26.3	29.0	34.7	32.8	31.9
No	67.7	63.8	56.1	52.0	60.2	58.5
Don't know/no response	12.5	10.0	14.9	13.3	7.0	9.6
No.	(353)	(320)	(269)	(248)	(201)	(188)

8 *What kind of Internet connection do you have at home? (Only for those who are connected to the Internet at home)*

	December 2001
Telephone line (normal)	74.2
RDSL	2.9
ADSL	7.8
Cable (Menta ...)	6.5
Others	0.5
Don't know/no response	8.1
No.	(383)

9a Are you personally in the habit of using the computer?

	June 1999	December 1999	June 2000	December 2000	June 2001	December 2001
Yes	40.7	41.2	44.9	47.3	49.6	49.7
No	58.8	58.7	54.8	52.7	50.4	50.3
Don't know/no response	0.5	0.1	0.3	0.0	0.0	0.0
No.	(1,000)	(1,000)	(1,000)	(1,000)	(1,000)	(1,000)

9b *Are you personally in the habit of using the Internet?**

	March 1998	September 1998	June 1999	December 1999	June 2000	December 2000	June 2001	December 2001
Yes	23.7	25.7	19.3	22.3	26.8	34.2	39.4	39.7
No	76.2	74.1	80.2	77.7	72.3	65.8	60.6	60.3
Don't know/no response	0.1	0.2	0.5	0.0	0.9	0.0	0.0	0.0
No.	(1,000)	(1,000)	(1,000)	(1,000)	(1,000)	(1,000)	(1,000)	(1,000)

* In March 1998 the question was: Have you ever connected to the Internet?

9c *Are you personally in the habit of using e-mail?*

	June 1999	December 1999	June 2000	December 2000	June 2001	December 2001
Yes	18.1	20.1	22.7	28.8	36.6	35.5
No	81.4	79.9	76.2	71.1	63.4	64.4
Don't know/no response	0.5	0.0	1.1	0.1	0.0	0.1
No.	(1,000)	(1,000)	(1,000)	(1,000)	(1,000)	(1,000)

10a From which place or places do you normally use the computer? (Multiple choice) (Only for those who use the computer)

	June 1999	December 1999	June 2000	December 2000	June 2001	December 2001
Home	75.4	72.6	79.3	79.9	85.1	80.1
Work	38.6	47.9	45.4	44.8	45.6	43.9
School	12.5	14.4	15.1	13.1	12.3	11.7
Others	2.2	3.7	2.0	6.1	3.2	7.2
Don't know/no response	0.2	0.0	0.2	0.0	0.0	0.0
No.	(407)	(409)	(449)	(473)	(496)	(497)

10b From which place or places do you normally use the Internet? (Multiple choice) (Only for those who use the Internet)

	June 1999	December 1999	June 2000	December 2000	June 2001	December 2001
Home	51.3	56.5	66.4	65.2	77.7	70.0
Work	40.9	45.3	46.6	43.0	42.6	40.1
School/Place of study	19.7	21.1	20.5	16.7	14.7	14.1
Others	4.7	4.9	2.6	8.2	3.6	9.8
Don't know/no response	0.0	0.0	0.0	0.3	0.0	0.3
No.	(193)	(223)	(268)	(342)	(394)	(397)

10c From which place or places do you normally use the e-mail? (Multiple choice) (Only for those who use the e-mail))

	June 1999	December 1999	June 2000	December 2000	June 2001	December 2001
Home	52.5	55.2	67.4	65.6	77.0	71.0
Work	45.3	51.7	49.8	49.0	44.3	43.1
School	16.6	16.9	17.2	14.6	14.5	13.5
Others	2.8	3.0	1.8	5.9	3.6	8.2
Don't know/no response	0.0	0.0	0.0	0.0	0.0	0.3
No.	(181)	(201)	(227)	(288)	(366)	(355)

11 How frequently do you connect to the Internet at home?* (Only for those who use the Internet or the E-mail)*

	June 2001	December 2001
Everyday	45.0	52.0
Weekly	33.5	34.7
Monthly	4.0	6.9
Less frequently	5.4	5.4
Never	10.7	0.5
Don't know	1.3	0.5
No.	(373)	(404)

* In June 2001 this question was only asked of people who connected from home.

Source: http://www.bcn.es/english/estadist/itaules.htm; updated 4 February 2002.

Chapter 4

The Case of Cape Town

1 Introduction

In this chapter, we will describe and analyse Cape Town's e-governance strategies. We will start in section 2 with a general context description of the enormous challenges that the City of Cape Town faces. This puts the city's e-governance efforts into perspective. In section 3, we summarise the city's information, communication and technology (ICT) strategy. In sections 4, 5 and 6 we analyse the issues of content, access and infrastructure respectively. Section 7 concludes.

2 Context

This section sketches the current economic and administrative conditions in the city of Cape Town, and lists the main challenges the city faces. This helps to understand the context in which the city pursues its ICT strategies.

Economy

The City of Cape Town is situated in the southwestern part of the Republic of South Africa. The metropolitan area has a population of around 3m people. The city accounts for 10.5 per cent of national GDP; it dominates the province of Western Cape, constituting 75 per cent of its economy (City of Cape Town, 2001). In the last two decades, the economy of the city has grown considerably faster than the national average: see Table 4.1. This trend continued in 2001, with Cape Town's economy growing by 3 per cent compared to the national economy, which grew by 2.2 per cent. In 2001, Cape Town's unemployment rate was 18 per cent.

The economic structure of the city is strongly biased towards the tertiary sector (see Figure 4.1). In terms of economic output, the trade and catering (including tourism) and financial and business sectors have grown fastest in the last few decades, from 35 per cent in 1980 to 42 per cent in 2000. The tourism sector in particular has grown tremendously over the past decade. Apart from these formal sectors, the city has a large informal economy. It is estimated that the informal sector employs more than 1,890 of the labour force and produces about 12 per cent of economic output (City of Cape Town, 2001a, p. 5).

Table 4.1   Percentage economic growth in Cape Town and South Africa, 1980–2000

	Cape Town	South Africa
1980–1991	2.2	1.6
1991–1996	2.5	1.9
1996–2000	2.5	2.1

Source: City of Cape Town (2001a).

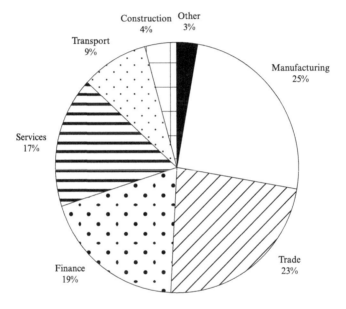

Figure 4.1   Sectoral division of real gross regional product of Cape Town, 2000

Source: City of Cape Town (2001b).

Administration

Politically, at the time of writing the city is ruled by the Democratic Alliance. This makes it the only South African city in which the African National Congress (ANC – the ruling party at the national level) does not have a majority. An executive committee runs the city. Unlike many other cities (such as Johannesburg) where an executive mayor system is in place, in Cape Town responsibility is shared among Executive committee members. Both ANC and DA members are Executive Councillors on the Executive Council.

The city has a total budget of R9bn (€750m), and an operating budget of R7.5bn (€630m). Some 40 per cent of this budget is spent on personnel. The city raises its income from rates/service charged and RSC levies. Only 0.5 per cent of the city's budget is provided by the national government. This gives the city a high degree of budgetary freedom, but also a large responsibility to pursue prudent financial policies. The city employs 28,000 staff, of whom 2,000 are temporary contract workers.

The local administration of the region has gone through major changes in the last decade. Up until 1995 there were 39 different councils in the metropolitan area. In 1996 this number was reduced to six local councils, plus one central metropolitan council for the coordination of the entire metropolitan area. In 2000, a single metropolitan government body was established for the entire Cape Town conurbation, which initially was called the Unicity but is correctly called the City of Cape Town. This decision was taken unilaterally at central government level. One of the key reasons was to create greater equality. Unlike in the past, rich enclaves and very poor neighbourhoods now fall under the same authority. In this new situation, public expenditures are more equally spread over the population and equality in services provision can also be reached. The mergers created enormous internal dynamics in the municipal bureaucracy and called for strong action to be taken to integrate the formerly independent units. For instance, there is a lack of common service delivery standards across the different areas in the Unicity and a lack of integration between departments and directorates. There is also a risk of increasing bureaucracy as the scale of the city has grown so much.

On top of that, the workload of the municipality tends to increase. On the one hand, in South Africa more and more functions are being devolved to local management. This is partly due to fiscal and budgetary constraints at the national and provincial level and often leads to the delegation of 'unfunded mandates'. Furthermore, there is a strong pressure to improve the quality of service delivery. The national government's 'people first' White Paper strongly calls on local governments to improve their performance and become more customer-oriented. Other stakeholders, like communities and the private sector, are also exerting pressure for better service and information provision. Given the scarcity of resources, all this implies that the city has to be creative in finding methods to mobilize capacity and resources outside the municipality for the development of the area.

From our interviews, we found the quality of management is good. We got signals that many excellent people in key positions in the organization were attracted from the private sector. Also, we found strong commitment to the public case and willingness to transcend the borders of people's own departments and think in a more integrative way. For the near future, one of the challenges of the city management will be to attract more talented blacks, in order to align the constitution of the civil officers corps with that of the population. Below the first reporting level of the organization, a 60 per cent target for appointments from employment equity designation groups has been set.

Key Challenges

The city faces a number of challenges. Unemployment is growing, as a result of economic restructuring and low skills levels of large groups. Poverty is a big problem and is expected to deepen and spread amongst a larger section of the population. At the time of writing, 53 per cent of the black and coloured population live just above subsistence level. Income inequality is rising, as job growth mainly takes place in highly-skilled jobs. AIDS/HIV infection levels and growth rates are reaching alarming levels. This specifically affects the poor and vulnerable people.

One of the key challenges of the city management is how to deal with meeting the housing and service needs of its population. The economic growth rate for the city was 3.2 per cent for the period 1996–2001 and this is projected to drop to 2.1 per cent per annum for the period 2001–2006.[1] For the city, providing housing and services to the newcomers are key challenges. Another key challenge is how to close the skills gap. As in other cities in the country, this gap is huge. On the one hand, there is a thriving part of the economy with highly-skilled jobs and high incomes – this is the advanced urban service sector. At the other end of the spectrum, there are a number of low-skilled (and low-paid) jobs, mainly in the social services and tourism sectors. However, unemployment is substantial and was even expected to rise from 18 per cent in 2000 to 21 per cent in 2005 (City of Cape Town, 2001a). One of the key reasons is that too many people lack even basic skills. To a large extent, this is due to the Apartheid educational system. The standard of Black schools was not high given that they received relatively limited resources and even now most do not meet the required standards. The skills gap trend is expected to increase unless it is proactively addressed. Adult illiteracy has grown by about 17,000 adults per year since 1996 and now totals 350,000 adults (City of Cape Town, 2001a).

Reducing the skills gap is seen as a priority, both from a social and safety perspective and to boost the local economy. Promoting tourism is regarded as strategic, as this sector absorbs so many relatively low-skilled jobs and also results in the promotion of the city of trade and investment purposes. However, even with the most optimistic projection, the growth of tourism will not bring jobs for everyone. The city also wants to upgrade the overall skills level so as to meaningfully participate in the emerging 'knowledge economy'. Education is regarded as the key instrument: the city has no formal competence in education

---

[1] The urbanization tendency can also be observed in other main South African cities. It has its roots in Apartheid. Under Apartheid, black people were not allowed to move into Cape Town, but since 1994 – when democracy and racial equality were established – an increasing number of people seek their fortunes in the urban area.

– this is a role of the national and province spheres of government – but it tries to achieve things informally, by maintaining good relations and alerting education institutions of the city economy's needs. For instance, it seeks to influence educational programmes of the local universities and polytechnics, and steer them in a direction that is in line with the strategic objectives of the city. *Inter alia*, the city is also supporting efforts to promote ICT education and thus contribute to the attractiveness of Cape Town for the ICT sector. In similar vein, the city is trying to encourage private firms to provide bursaries in the field of ICT studies.

Safety and violence are major challenges in Cape Town, as in all other South African cities. Limited prospects for jobs and social disintegration are a fertile soil for violence. Crime rates are high, although by South African standards the safety level in the city is reasonable. In the longer run, fighting crime is largely a matter of creating better economic and social prospects for large groups of the population. In the short run, the burden is on the shoulders of the police force. Recently a municipal police corps was established to complement the national police service. This puts a heavy burden on the city's budget, but has started to make a difference to safety levels in the city itself and in other targeted suburbs.

In conclusion, urban management in Cape Town has to face dramatic internal and external challenges. As we shall see later in this chapter, information technology is seen as a strategic and enabling tool to begin to address a number of them: to help close the skills gap, to improve the quality of urban management, to increase the efficiency of municipal services delivery, but also to strengthen the local economy and reduce social exclusion. All the issues are included in the city's 'Smart City' vision, which will be discussed next.

3   Cape Town's and Western Cape Province's ICT Vision and Strategy

Introduction

In this section, we start our analysis of Cape Town's e-governance performance. We will first take a look at the city's ambitions, vision and strategy. Next we review the strategy of the province in this respect and finally in this section we sketch the national vision on ICT deployment.

Cape Town's Smart City Vision

The city has elaborated a vision in which it expresses what the city wants to be. The key ambition of the city is to be a safe city that is clean, attractive to investors, welcoming to visitors and underpinned by a vibrant, growing economy. Furthermore, Cape Town wants to be:

- a well-run democratic city, accountable to the people, corruption free, transparent in its activities and prudent in its financial management;

- a city in which no one is left behind and where everyone has equal opportunities and is guaranteed basic services;
- an open, tolerant city;
- a city filled with concerned and responsible citizens;
- a city which is the best place to live, work, invest and visit in all of South Africa.

Finally, and very relevant for this case study, Cape Town wants to be a *smart city, populated by informed people, connected to the world and each other by the technology of the information age* (City of Cape Town, 2000).

The city wants to achieve these ambitions in a partnership with its entire population. It is convinced that Cape Town's strong points – its spectacular natural beauty, rich heritage and mix of cultures – should be built upon.

Thus, Cape Town is the only city in South Africa where information technology forms an integrative part of the city vision. ICT is embedded in a number of policy fields. It is seen as an instrument to foster the economic development of the city. This translates into strategies to promote the local ICT cluster, to steer local education programmes in such a way that the educational offer of ICT courses will improve and increase and to help people develop IT skills and find their way into or back to the labour market. ICT is also regarded as an instrument for social policy. This is reflected in the commissioning of digital divide research, aiming to ensure access for all citizens as well as other social initiatives (see section 6 for elaboration). Next, ICT is used as a crucial tool to enable good governance. This is reflected in substantial investments in information systems in the municipality (see section 5). Ultimately, this should lead to more effective and efficient service provision.

In its Smart City Strategy, the city distinguishes five focus areas:

1   *Create smart city Leadership.* To achieve its goals, the executive and senior levels in the organization should be aware and convinced of the possible contribution of ICT to urban development issues. To create awareness at the highest political level on the potential of ICT as instrument for urban management, all the Councillors were equipped with personal computers, access to the Internet and training. This created more support for integration of ICT into Council policies and strategies. Councillors recognize the value added of computers and now e-mail and the web are widely used to communicate and exchange documents. Minutes of council meetings are also published on the Internet.

2   *Create a favourable policy and regulatory environment.* To do this, the city is in the process of assessing whether its current policies are 'digital age proof'. Frameworks in the field of privacy, data protection, service levels, etc. are being developed. Finally, the city seeks to influence national policy making in the fields of ICT and e-government, as many strategic decisions and frameworks which are relevant for the city are taken and developed by this sphere of

government. For instance, city officials participated in the drafting of the national White Paper on e-commerce.

3 *Implement an e-government strategy.* The city introduces ICT systems not as an aim in itself, neither to automate existing processes, but rather to re-engineer local government to improve its performance and lower its costs. The implementation of the SAP Enterprise Resource Planning system is currently taking place (more on this in the next section). The Unicity Commission, charged with the coordination planning and preparing for the seven previous Local Councils into one city, recognized the need for simultaneous e-government efforts in a very early stage (2000). Currently, systems and processes that will support e-government are being introduced side by side with the citywide transformation process. The momentum of change is optimally used to upgrade the city's IT capacities.

4 *Use ICT as a tool for economic and social development of the city.* As mentioned above, this includes supporting and developing the IT sector, promoting the development of IT skills among the population, promoting ICT access and education. Furthermore, the city plans to improve its data collection and management processes which will ultimately ensure more informed and coordinated planning and targeted interventions.

5 *Improve governance/digital democracy.* ICT is used to make local government more accessible, transparent and accountable.

The objectives mentioned above and the programme list indicate that ICT is an integral part of the city's developmental strategy. Another example of the embeddedness of ICT in more general strategies is shown in the Joint Marketing Initiative (JMI). ICT, its use and access is central to the city's strategic economic development objectives of building competiveness and providing a business and visitor friendly environment. The JMI is an initiative by the Province of the Western Cape and the City of Cape Town to focus and align the marketing efforts of the various tourism organizations, investment agents and other marketing ventures, in the Western Cape and Cape Town.

Western Cape's Vision and Strategy

The Province of Western Cape, in which Cape Town is situated, regards information technology as a strategic tool to achieve economic, social and educational goals. As to the economic part, in a recent White Paper, the provincial government of Western Cape has committed itself to preparing the Western Cape for the knowledge economy of the twenty-first century. It realizes that 'the ability to maximize the use of knowledge is now considered to be the single most important factor in deciding the competitiveness of regions as well as their ability to empower their citizens through enhanced access to information' (Provincial Government of the Western Cape, 2001, p. 4).

In its Cape Online Programme, the province outlines its vision on how ICT fits in its knowledge-oriented strategy. The *vision* of the Cape Online Programme is: 'To develop an innovative environment that facilitates a competitive knowledge based economy that promotes economic growth and enhances the quality of life for our people' (p. 11). The *mission* is defined as 'enabling government to harness the capabilities of the Internet, to grow the appropriate use of ICT, increase internal efficiencies and provide a better services to its citizens as a pathway to e-Government'. Five policy objectives (p. 13) were formulated:

- raise awareness on the vital role of ICTs in the economy;
- develop and expand ICT infrastructure to world-class standard;
- use the ICT revolution to enable everyone to access and use information to maximum benefit and to conduct business more cost-effectively;
- use ICT to improve the efficient and effective delivery of government services;
- promote cooperation and collaboration in the development and operation of ICT between all sectors.

Thus, Cape Online has internal and external goals. The internal goals are directed towards improving the efficiency and effectiveness of the provincial government itself and improving the level of services delivery. The external goals are directed towards improving the quality of the infrastructure, increasing the use of ICTs in business and communities and promoting ICT access and skills. The objectives are translated into various projects. Three types of project are discerned: core projects (impacting the core role of the government); online-community projects (impacting various communities of interest, specific groups of citizens or companies); and external projects (non-governmental projects that impact the online environment of businesses, organizations and individuals) (see Figure 4.2).

We will describe some of the projects in more detail in the sections 4, 5 and 6, under the headings of content, access and infrastructure policies respectively.

South African National ICT Vision and Strategy

The national government of South Africa strongly supports and encourages the countries' transition towards a knowledge economy. It realizes that the deployment of information and communications technologies is an important element in this.

National government is an important player in the ICT policy fields. In the first place, it is responsible for the overall conditions in which the transition to the digital economy can take place. It sets standards and legislation for digital identification, defines property rights and sets telecom legislation. Second, national government promotes the creation and use of ICTs throughout society

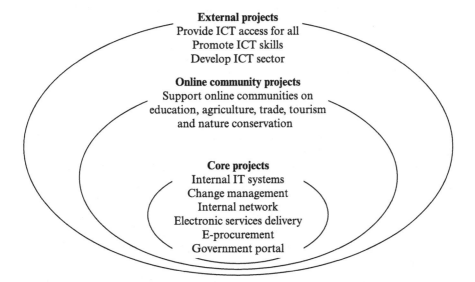

Figure 4.2 Facets of the Cape Online programme

in various ways. Third, national government is an important ICT user itself, and has ambitious plans to introduce e-government in its own organizations.

A number of national government departments have issued white paper and other policy documents that elaborate strategies for the adoption of ICTs in society. Some of them are: the e-commerce Green Paper published by the Department of Communications, electronic government, the Digital Future IT policy framework by the Department of Public Service and Administration (http://www.dpsa.gov.za), and the White Paper on Science and Technology by the Department of Arts, Culture, Science and Technology (http://www.gov.za/whitepaper/1997/sc&tecwp.htm).

4 Governing Content

This section describes and analyses the creation of electronic content by the city and the province. The main focus is on the introduction of internal IT systems, and the web strategy of the city.

Internal ICT Systems

The current 'Unicity' of Cape Town was established in 2000 as a consolidation of seven municipalities. In the early 1990s, the number of municipal authorities was even greater. This legacy is reflected in the current IT context in the city. In terms of IT, the city finds itself using a multitude of IT systems: every municipality, and almost each department, has its own hardware, networks and applications. Some observers estimate that a staggering 270 systems are currently in place. This situation seriously hampers the management of the Unicity. Systems do not communicate well, so the information provision does not meet the demands of even basic management standards.

Both the City Manager and the newly-created centralized IT department were strongly convinced that a new, integrated information management system needed to be put in place in order to streamline information flows and integrate all the different systems. It was decided to put an integrative Enterprise Resource Planning (ERP) system in place to meet these requirements. After a thorough and rigorous selection procedure, the city selected SAP as the platform and a consortium led by Accenture as consultant to implement the system. The first phase of the system incorporates and integrates human resources and payroll systems, financial accounting and management accounting, real estate management, revenue management, customer care and materials management/procurement. It will replace or integrate at least 150 of the current systems.

The implementation of an integrated system of this scale and complexity is fraught with challenges and has major impacts on the entire municipal organization. To ensure the successful implementation of the project, a collaborative partnership was established between Accenture and the City of Cape Town. In the first stage, all the top managers of the departments were informed about SAP and its characteristics. Next a systematic process of implementation was started, in which reorganization and system implementation went hand-in-hand. This is because in many cases systems integration asks for new ways of working. In total, 250 people were involved in the implementation, of which 125 were from Accenture and 125 from the city. It is a long-term project.

Introducing such a system is costly: in total, R285m will be invested. However, significant returns can be expected when the system is in place. Firstly, there are direct monetary gains. It will improve the city's capacity to collect taxes and revenues from utilities (electricity, water, solid waste collection). At the time of writing there is a significant tax revenue gap as a result of poor quality data and management systems. Secondly, there are intangible benefits. Most importantly, management information is likely to improve substantially, with better policies and decision-making as a probable result. For instance, the new system will enable the assessment and monitoring of the value of municipal property in a consistent manner, something that is currently not possible. Furthermore, the quality of service delivery is likely to improve as well.

The size and reach of this project is substantial: the speed at which it has been implemented is impressive. One of the reasons seems to be the city's strong IT department. Not only did they manage to mobilize substantial resources (both finance and people) to implement the system, they also convinced the Councillors of the need to get things done. This should be regarded as a positive achievement in a city where there is competition for scarce resources.

It is interesting and curious that so many people with ICT skills apparently still want to work for the municipality, whereas they can earn high salaries in the private sector. Many of them find it challenging to be involved in such a major scale ERP introduction.

Apart from the ERP introduction, there is another city initiative around the creation of standards. At the time of writing, each department has its own office hardware and software applications. This situation has to change: the city wants uniform standards in this respect also. As for hardware, the city wants a single standard from one large supplier. The city's approach to the facility management (maintenance, application management, infrastructure) is to hire several local small business firms, to let them benefit from the new initiative also. Hiring small, local firms bears a certain risk, but at the same time it can be a tool to boost the local ICT sector and develop new IT capability in Cape Town. The city needs to balance the risks against the benefits carefully.

To control and coordinate the overall IT function, the city might engage in a partnership with an IT company. The starting point is that whatever is outsourced, the city will be in the driver's seat on a strategic level.

Like the City of Cape Town, the province of Western Cape wants to improve its IT capacity to improve efficiency and services delivery. In the 'Cape Gateway' project the province seeks to identify and fill process gaps in the organization and optimize the administration's information infrastructure. The 'Cape Change' project is a management project that identifies aspects in the provincial government's processes, procedures and mechanisms that will need to change to bring e-government into effect. Workflows, work processes and job descriptions will be redesigned.

Web Strategies

Cape Town's web activities reflect the legacy of the history of the organization. At the time of writing there are seven intranets, one for each former Local Council. Through a natural process the seven intranets should merge into one. Already, the IT department has started a 'City Web', including a discussion forum for civil officers, a second-hand virtual marketplace, and an online phone directory. The IT department coordinates the web development of the entire municipal organization.

One of the problems concerning web content is that the individual departments did not feel responsible for their web content: their responsibility ended with submitting news and text to the IT department for publication on the web. To increase web awareness and responsibility, the IT department has provided each of the municipal departments with a web content management tool. This enables and encourages officials to maintain their own websites and makes them responsible for the content. The risk of such an approach is the supply-oriented nature of the content that is likely to be published. In our model, demand orientation should be the guiding principle. In this case, internal demand for content should be established and the web content should be structured along that line. There should be a coordinator on the type and format of content that is needed from each department. Currently, there is no such 'strategic' integrator of content. The IT department only functions as technical integrator.

The city's Internet site is also posted and maintained by the IT department. As one of the features, people can put questions to councillors and check whether their questions are reflected in council debates (the minutes are published as well). However, in general, at the time of writing, the content of the Internet site is poor and not demand-oriented.

The city seems to opt for an 'inside out' strategy: first organize the internal information provision and then use the Internet as means to communicate with other stakeholders. The implementation of the SAP-ERP solution is the first step: when implemented it will 'power' a wide range of interactive possibilities via Internet. It is a trigger for the Internet strategy. A complicating factor is the role of the communications departments. Each of the former seven Local Councils had its own communications department: some of them had developed their own website. The challenge will be to involve these people in the development of the Internet strategy. In our view, in the medium term, an integrative, demand-oriented Internet strategy is needed.

The province of Western Cape plays an active role in web development: it has initiated a provincial portal website, 'Cape-on-Line'. This portal should connect 450 websites of public organizations. The idea is to enhance transparency of the web-offer for individual citizens by offering online access to all public content. The Province also promotes the formation of virtual communities of interest groups, e.g. farmers.

City and the province activities seem well connected. Cape Town city officers are on the board and steering group of the Cape Online initiative.

All Cape Town public tenders are published on the Internet. This makes procurement more transparent and provides equal chances to all subscribers.

5   Governing Access

Access to ICTs is a pressing issue, particularly in South Africa. According to a CIA estimate, in the year 2000 there were 1.82m Internet users in South Africa (http://www.cia.gov/cia/publications/factbook/index.html). This amounts to 4 per cent of the South African population. By Western European standards this is a low figure. However, growth is fast: a more recent study by BMI-TechKnowledge has put the number of Internet Users in South Africa at 2.6m in 2001 (6 per cent of total population), with 1.6m devices being used to access the Internet. Although there are no regional data available, the Provincial Government of the Western Cape (2001) estimated that Cape Town and the Western Cape region, together with the Gauteng province, were frontrunners in South Africa.

In this section, we will discuss ICT access policies of Cape Town and the Province of Western Cape. Both feel the need to improve levels of ICT access, not only among the population, smaller companies and communities, but also within their own organization. For instance, only 7 per cent of provincial employees have access to e-mail. Both the city and the province are convinced that improving access brings many benefits. Improving ICT skills among the population is one of the priorities. This improves employment opportunities – basic ICT skills are required for many jobs – but also enables people to benefit from the many conveniences that ICT brings. For instance, ICTs can bring medical applications to remote villages, offer better educational tools and give local business access to global markets.

In Cape Town we found several innovative programmes to bring the benefits of the digital era to disadvantaged communities. One initiative brings computer facilities into schools. They are not only meant for the schoolchildren but the terminals are also accessible after school hours and at the weekends. Schools in poor communities have no money to buy computers and connect them to the Internet and also lack capacity to use and maintain the systems. Many children remain deprived of even basic computer skills. The project takes a grassroots approach, closely involving local communities in the programme. This ensures sufficient levels of trust among communities and greatly increases efficiency. An Internet service provider is part of the project. This company wanted to do something for disadvantaged groups and now provides the schools with connections (for two years), computers and applications. An important aspect of the project concerns the build-up of skills. Schoolteachers are trained to work with the computers and to teach the children and adults how to do it.

One of Cape Town's main aims is to enable every citizen to benefit from the Information Revolution. In this context, at the time of writing a digital divide assessment (the first to be undertaken by a city worldwide) is underway to establish the needs for ICT among various groups of the population in the metropolitan area and the constraints and opportunities to improve access to ICT. The study also includes a review of the existing ICT initiatives being undertaken in the city by a range of organizations at the time. The needs assessment will be used as input to direct ICT policies and programmes to specific targets groups. It will also guide the efforts of companies that want to invest in bridging the digital divide. If the city administration knows where and what the needs are, it is much easier to assist companies to target their social investments to the areas where the needs and benefits are greatest.

Private companies in many cases are willing to contribute to regeneration in poor neighbourhoods from a corporate social responsibility perspective. These companies invest in computers and Internet connections. Microsoft, for instance, offers charge-based Internet access in six locations in the city. There are indications, however, that in too many cases private benefactors just 'dump' the money and run away. In this way, very limited structural improvements can be made. Often there is insufficient trust among the local target population that the company is really willing to improve things in the neighbourhoods; in many cases, companies understand the nature of the needs of the population insufficiently.

In another programme, public libraries – owned by the city – are being connected to Internet. This project is also linked to a skills development programme: unemployed people are trained to maintain computer networks in libraries. This working experience may ultimately result in a regular job. A further role of certain libraries which have 'Business Corners' is to provide information to SMEs on general issues and on the opportunities of ICTs for their business operations. The libraries also serve as contact point with the municipality (that has become very large now, with greater distance to the individual citizens). In an interesting and innovative project, the City of Cape Town links the libraries with nearby schools, thereby providing Internet access in those schools as well.[2] In its 'Cape Online' programme, the province of Western Cape aims to equip all Western Cape schools with a full educational and administrative IT system (Provincial Government of Western Cape, 2001). The use of technology will be integrated into current curricula. Finally, the province has also established Internet kiosks in some disadvantaged communities.

6   Governing Infrastructure

This section on electronic infrastructure is a short review. Cape Town is said to be the 'best wired' city in Africa. Nevertheless, from an international perspective the IT infrastructure is poor. This is mainly due to the monopoly position of Telkom, the incumbent telecom operator. Prices are relatively high and quality of services is low. Under current legislation no company can construct its own cabling between two buildings, unless they are adjacent. In all other cases, you have to lease a Telkom line. The average bandwidth is 2Mb: higher bandwidth is extremely expensive.

Cape Town's ambition is to become a world-class city with a strong ICT profile. Reform in the telecom situation in South Africa is a precondition to realizing this ambition, as infrastructure is one of the basic provisions on which a true knowledge economy thrives. As telecom legislation is a national issue, Cape Town should do everything in its capacity to influence national policies into a more competitive market. In doing so, it could very well cooperate with other major South African cities with similar problems and ambitions.

7   Conclusions

In this chapter we have analysed Cape Town's ICT strategies. As the Province of Western Cape is also a very active and influential player in the various ICT

---

[2]   Schools are the responsibility of the Province not the City.

policy fields, we have also paid generous attention to its ambitions and activities in this respect.

It is clear that Cape Town is a city confronted with considerable challenges, particularly compared to the European cities that we also analysed. One of the key challenges is to close the skills gap – of which the 'digital divide' is a derivative – to improve the quality of life for the growing number of inhabitants and to guide the city into the knowledge economy. On the whole there is a tension between developing the advanced part of the economy (by promoting future growth industries and advanced technologies) and fighting poor living conditions for large parts of the population. Another challenge is to manage the large unified city in a comprehensive way. Information technology policy is at the intersection of these three lines. It can contribute to employment opportunities and quality of life of many people; it can help to build the knowledge economy; and it is a tool to improve urban governance.

In our study we have developed a framework of reference in which we make a distinction between three 'local dimensions' of ICTs: content, access and infrastructure. There are strong indications that the three local manifestations of the information society are interdependent and sometimes mutually reinforcing. We suggest that its dynamics can be represented as a local 'digital flywheel', which functions as follows. If there are more ICT users (access) in a city, it becomes more interesting for companies or any other actors to develop new services (content). For instance, online grocers normally start their activities in areas where Internet penetration is highest. On the other hand, more (or better) electronic services (content) may increase the number of local users. If there are better online products or services available, the Internet becomes more useful and more people are likely use it. This interdependence between access and content is well known in the economic literature on technology adoption (see Leighton, 2001). In many instances, a 'killer application' can speed up the adoption of a new technology very rapidly. The quality of local content is probably not the key factor for individual's decisions to buy a computer and go to the Internet. Nevertheless, several studies suggest that local information and services are very important to citizens (Anttiroiko, 1998; Servon and Nelson, 2001; Baines, 2002).

For cities, 'turning the flywheel on' may bring benefits in several respects. Improved electronic services mean a higher quality of life for inhabitants: they have better access to improved amenities. E-government services may save public spending and reduce local taxes to the benefit of citizens and/or firms. Also, e-government may improve local decision making as is improved the quality of management information. The quality of local electronic infrastructure is a factor of growing importance to attract or retain inhabitants (Healey and Baker, 2001). Wired homes have the potential for being seen as more upmarket and desirable than others (Baines, 2002). Virtual communities can contribute to safety, social cohesion and political participation (van Winden, 2001). High-quality infrastructure is also important to attract to or retain firms in the region. Furthermore, policies may bring 'first mover advantages'. If a region manages to create early mass in users and infrastructure, local firms may build an innovative edge. In particular, early critical mass of users may attract innovative companies and people into the city. The system takes off when a critical mass of users is reached.

A question that comes to mind is how local is the 'local flywheel' really? Clearly, its engine is not solely fuelled by local factors. External factors play an important role, too. In the first place, national institutional conditions matter. Our South African case study suggests that it makes a big difference whether the telecom market is liberalized and competitive or not. All other kinds of legislation influence the 'flywheel' as well – for instance, electronic privacy and security legislation. Second, general economic conditions play a role. ICT use is strongly related to economic development levels. Richer countries and cities tend to have higher levels of access, more content to offer and a higher quality of infrastructure. Third, national policies can strongly influence the different parts of the 'flywheel'. Regarding access, many countries have nationwide programmes for ICT in education or access policies for disadvantaged communities. In the field of content, national policies may encourage cities or other public entities to develop e-strategies,and thus speed up the quantity and quality of content offered. Despite all this, our study has revealed that there is still sufficient scope for urban policy makers to do something.

What is Cape Town doing to turn the local 'digital flywheel' on? Our analysis of Cape Town's ICT strategies suggests that the city has a high strategic level of thinking and acting. Furthermore, ICT strategies are very well embedded in several other city strategies.

The e-visions and strategies of both Cape Town and the Province reflect a comprehensive approach, harnessing ICT both in policy goals (social, economic) and processes (re-organizing the municipality and the province to make them more effective and efficient). Regarding content, Cape Town scores highly on its internal capacity to introduce a comprehensive information system (SAP) that will improve governing capacity and improve service delivery to residents.

In the near future, the implementation of the ERP system in Cape Town offers a number of opportunities for information and (interactive) services provision to urban stakeholders. Citizens will be better informed, procurement can be more cost-effective and revenues may rise as tax collection will improve. More generally too, the system will improve management information provision and improve urban governance. The Province's plan to create a government portal will improve the accessibility of local and provincial government for individual citizens and companies.

The quality of local content is an important driver of ICT use: content creates the value added of the technology. An interesting and promising project in this respect is the Province's activity to support and promote the formation of online communities of special interest groups. This project capitalizes on the Internet's unique features.

At the time of writing, web content for citizens, businesses, tourists and other stakeholders is rather poor and fragmented. The value of the Internet as a key instrument for city marketing is not capitalized on. For tourism particularly – one of Cape Town's growth sectors which offers employment for low-skilled people – the degree of 'Internet organization' is rather low. We are critical of the external presentation of Cape Town on the Internet. There seems to be no strategy in this respect. We feel that a demand-oriented web strategy is needed, one that addresses the needs of various 'urban user groups'. Such a strategy requires not only reliable 'public' information from the city, but also from many other actors such as the local tourist sector, transportation companies and many more. An integrative web content management approach is needed: given the nature of this task, the IT department is not the appropriate entity to run this. There are several options open. Some cities have developed (or invested in) comprehensive 'urban portals' as a preliminary web page, combining public and private content organized around user profiles. Other sites are organized around life events (marriage, moving house). The City Management has to answer the question of which role it wants to play in providing and organizing information on the web. One minimal option is to provide only public information and services to citizens and firms and let private content suppliers organize themselves. Another option is to take a more active role in promoting the organization of content in subsectors such as tourism, or create an urban portal with links to various relevant sites.

In the field of access, the city is very well aware of the need for access to information technology and the Internet (in terms of availability and skills). More than in European cities, access policy is urgent: only a small proportion of the population has access to the Internet and many people lack even basic ICT skills. The city, and the Province also, have initiated a range of programmes throughout the city. There is a strong belief that ICT can contribute to economic and social regeneration, particularly among the poorest groups in the city, and that the city can play a positive role both as initiator of projects and also as enabler or facilitator of private initiatives. The digital divide study underway is a unique project that will yield new insights into the 'digital map' of the city and will help to guide policies as well as private initiatives to reduce the divide. This approach can serve as best practice for many other cities. One problem for Cape Town with regard to access policies is its lack of direct influence on education, which is a provincial responsibility. This restricts the city's ability to improve ICT skills among the entire population.

Regarding electronic infrastructure, the situation in Cape Town is favourable when compared to other African or even South African cities. However, in terms of world standards the ICT infrastructure does not meet demands. The city is not to blame. Rather, national telecom regulation results in a monopoly position of the incumbent telecom supplier, prevents innovation and keeps prices relatively high and available bandwidth low. If Cape Town wants to be a world-class city, this situation needs to change. In particular, the ICT sector demands excellent ICT infrastructure conditions, but the same holds true for ICT-intensive industries (such as banks or other service industries). The city should do what it can – in cooperation with other cities – to change the current situation. In Europe, cities are becoming more active players in the domain of local electronic infrastructure. The Dutch city of Groningen, for instance, is developing a public optic-fibre network, linking all public buildings in the city. At a later stage companies are to be linked as well. By bundling demand it can also promote competition between telecom services providers, leading to lower prices and better services.

Our overall perception is that e-governance in the city is of high quality. We identify a number of drivers behind this power. Firstly, the sheer magnitude of the challenges that urban management is confronted with seems to trigger strategic thinking. The lack of resources compels the city management to think hard about the allocation of scarce resources; the result is the high quality of policy making. Secondly, officials in the municipality are used to working and experiencing an ever-changing environment. They are less 'entrenched' than people in many European cities where the dynamics and problems are less outspoken. This attitude greatly facilitates the introduction of e-government, as this requires lots of internal adaptations as well.

Recently, the City of Cape Town has become a full member of the newly-established South African Cities Network, consisting of South Africa's major cities. The network aims to provide a vehicle for the sharing of information and research between the nine major cities in the country. Regarding e-governance, this network opens new perspectives in several respects. It can serve as a platform to exchange views and experiences on how to develop e-government policies, how to increase levels of access, how to involve private companies in policy design, etc. The network could also be an instrument to put the city's interests higher on the national agenda. In this chapter we have first concluded that the quality of electronic infrastructure is such that it hampers cities' abilities to develop effectively into a 'knowledge society' and to attract ICT-reliant business. Cities that are particularly impacted by this may join forces in their lobby for better legislation, or take a seat on national commissions that deal with telecom issues. Secondly, we feel that cities in South Africa have relatively little influence on education, which is crucial in any 'smart city' or 'world class city' strategy. The major cities network could be a vehicle to find ways to increase cities' influence on education, but also provide a platform for knowledge and experience sharing and exchange in this respect.

References

Anttiroiko, A. (1998), 'Planting the Seeds of Economic Growth and Social Welfare: Local and Regional Governments in Finland and Korea Facing the Challenge of the Information Age', paper prepared for the International Conference on Electronic Democracy EDI, Korea.
Baines, S. (2002), 'Wired Cities', *Communications International*, April, pp. 21–25.

City of Cape Town (2000), 'Transforming Local Government with an IT-enabled Strategy – Cape Town's "Smart City" Strategy', presentation.

City of Cape Town (2001a), *Towards an Economic Development Strategy for the City of Cape Town*.

City of Cape Town (2001b), *Cape Town's Economy: Current Trends and Future Prospects*.

City of Cape Town (2002), *Key Economic Statistics Update – 2001*.

Healey and Baker Consultants (2001), *European E-locations Monitor*.

Het Financieele Dagblad (2001), 'Geschoolde blanken Zuid-Afrika wijken uit naar buitenland', 9 July 2001.

Leighton, W.A. (2001), *Broadband Deployment and the Digital Divide: A Primer*, Policy Analysis No. 410, August 7.

Provincial Government of the Western Cape (2001), 'The Cape Online Programme', version 3.9.

Servon, L.J. and Nelson, M.K. (2001), 'Community Technology Centers and the Urban Technology Gap', *International Journal of Urban and Regional Research* 25 (2), pp. 419–26.

van Winden, W. (2001), 'The End of Social Exclusion? On Information Technology Policy as a Key to Social Inclusion in Large European Cities', *Regional Studies* 35 (9), pp. 861–77.

Interview Partners

Andrew Boraine, Special Advisor to the Minister, Ministry of Provincial and Local Government.

Scott Fitzmaurice, Project Leader, Department of Information Technology, City of Cape Town.

David Gretton, Director, Department of Economic Development, Tourism and Property Management, City of Cape Town.

Mymoena Ismail, Department of Information Technology, City of Cape Town.

Alan Levin, Project Officer Cape Online, Section of Knowledge Economy & E-government, Province of Western Cape.

Nathan Momsen, Business Development Solutions.

Donovan Muller, Partner, Accenture Cape Town.

Mark Neville, CEO, Future Perfect Corporation.

Demitri Qually, Councillor, City of Cape Town.

Nirvesh Sooful, Director, Department of Information Technology, City of Cape Town

Belinda Walker, Deputy Mayor, City of Cape Town.

Harold Wesso, Head, Section of Knowledge Economy and E-government, Province of Western Cape.

Carol Wright, Interim Branch Manager, Knowledge Management Economic Development and Tourism Directorate, City of Cape Town.

Chapter 5

The Case of Eindhoven

1 Introduction

The city of Eindhoven proudly presents itself as 'leading in technology'. With many well-performing high-tech firms in the city and a technical university, its technological profile is indeed very strong. For the city's high-tech firms, information and communication technologies are important. For one thing, ICTs are increasingly part and parcel of the region's high-tech activity. Furthermore, and partly as a consequence, the city has developed as a centre of expertise in the field of ICT. ICT has become a key technology, not just for companies. The public sector in the city has also come to realize the importance of ICT, as a means to improving the quality and performance of the city in many respects. The city has initiated various projects to capitalize on the new possibilities of ICT, in cooperation with citizens, business and the local knowledge institutes.

In this case study, we will analyse Eindhoven's e-governance strategy. We will focus on two dimensions of e-governance. First, we will pay generous attention to Eindhoven's efforts to make Eindhoven a frontrunner in broadband, by connecting many of its households to fibre-optic networks. The city has elaborated a remarkable strategy, in cooperation with national government and private players, to achieve this. Our second focus is on the municipal electronic service delivery to its citizens. How is Eindhoven using the new technology to improve the quality of its public services? The two issues are interrelated: the city of Eindhoven wants to pioneer the provision of broadband services to its 'super-connected' citizens.

This chapter is structured as follows. Section 2 describes Eindhoven's e-strategies in general. Section 3 deals with local content policies, among which municipal electronic service delivery, Internet and broadband content promotion activities. Section 4 describes the city's access policy, and section 5 focuses on the infrastructure policies. Section 6 concludes, and looks at Eindhoven's e-strategies within the framework of analysis.

2   Eindhoven's ICT Strategies

Introduction

In this section we give a brief overview of Eindhoven's ICT strategies. First, we give a short background of the city's economic and technological profile. Next, we elaborate on Eindhoven's ICT strategies.

Profile of the City

At the beginning of the twentieth century the city of Eindhoven was no more than a small agricultural town with some 5,000 inhabitants. The foundation of Philips Gloeilampen NV in 1891 marked the beginning of the rapid development of the city (Adang and van Oorschot, 1996). Nowadays, Eindhoven is the fifth city of The Netherlands, with 198,000 inhabitants. It forms the centre of the region of Southeast Brabant, which is often referred to as the Greater Eindhoven Area, or the 'Eindhoven region'. This region comprises 34 municipalities. Like the city of Eindhoven, the population of this region has grown rapidly in the last century. Today, it counts some 670,000 inhabitants.

The appearance of the city of Eindhoven is determined by urban design of the twentieth century. In the city centre, buildings and housing estates offer a cross-section of twentieth-century (housing) architecture. Eindhoven's fast growth and its preoccupation with technology have also left their mark on the city's appearance. Although Eindhoven is the fifth city of The Netherlands, up to the 1980s the urban environment did not live up to that status. The town centre looked chaotic and apart from the Evoluon there were no special attractions. Since the 1980s, Eindhoven has put much effort into city renewal. A high quality shopping centre (Heuvelgalerie) and a concert hall (Muziekcentrum Frits Philips) have been opened, as well as a multifunctional building (the Witte Dame, a restructured large Philips building). Eindhoven is trying hard to bring its urban environment in line with its economic, social and cultural position.

Geographically, the region of Eindhoven is situated at some 100km west of the Randstad, the cultural and economic 'gravity centre' of The Netherlands. Nevertheless, its location is by no means peripheral, in particular in a European perspective. The region is situated within the rectangle formed by the Randstad, central Germany, the Ruhr Area and the Belgian cities of Brussels and Antwerp. The Eindhoven region ranks among the more prosperous in The Netherlands. Incomes per capita are well above the Dutch average. The region of Eindhoven is an important centre of employment. It counts some 307,000 jobs in a population of 670,000.

The economic fortunes of the region have changed several times in its recent history. Up to the mid-1980s the region did relatively well, with growth rates higher than the national average. From then on, however, regional employment growth was below the national average. Between 1986 and 1991, the yearly employment growth rates dropped from 5.3 to 0.9 per cent. From 1991 onwards, things got worse. On balance, the region started to lose jobs. Up to 1994 the regional employment figures continued to fall. The main factors behind the economic problems of the region were a reflection of severe difficulties of the two leading industrial firms in the region: Philips (electronics) and DAF (lorry construction). In the early 1980s Philips employed 35,000 people in the region. This figure had dropped to 21,000 by 1993. In 1993, the DAF company collapsed, meaning the loss of another 2,500 jobs. Additionally, the large network of external suppliers in the region that worked for DAF was also severely hit. After a few difficult years, the region recovered strongly in the mid-1990s. By many economic indicators the region shows higher growth rates than The Netherlands as a whole.

Along with a general recovery of the economy in the second half of the 1990s in The Netherlands, the severe restructuring of Philips and DAF in the beginning of the 1990s has generated positive spin-offs in the form of newly-started businesses and the outsourcing of activities.

The economic structure of the Eindhoven region is dominated by industry, but trade, business services and health and welfare services are also large sectors with regard to employment. The share of industry in the regional economic structure (in terms of both employment and added value) is one of the highest in The Netherlands. The region hosts some well-known industrial firms such as Philips Electronics and DAF as well as fast-growing high-tech companies such as ASM Lithography, Simac and Neways. The industry is very modern and knowledge-intensive. Compared to other important industrial regions in The Netherlands, Eindhoven's industry is very innovative: 50 per cent of total Dutch research and development (R&D) expenditure takes place within the region (Kusters and Minne, 1992).

The E-City Initiative

With the start of the 'e-city', the year 2000 marked Eindhoven's increased ambition level concerning e-policies. The e-city started as an initiative of the Dutch government, notably of the Ministry of Transportation and Telecommunication. The Dutch national government has an explicit strategy to speed up the transformation of the Dutch society into an information society. Several policy instruments are used, such as contributions to Electronic Highways, an integrative support scheme for ICT starters ('twinning'), the provision of Internet access for less favoured groups and many more. The e-city project is a recent innovative instrument. Every Dutch city had the opportunity to come up with a plan to design a 'knowledge neighbourhood' in the city. Several conditions had to be met: the area should host a mixed population in terms of income, race and social position; the project should be able to offer innovative services, and provide adequate infrastructure and access; public and private players should be involved and private investment in the area should be substantial; and (elements of) the project should have potential to be applied on a larger scale in due time. Also, the organization of the project should be very sound. Several Dutch cities submitted a project proposal. In July 2000, the State Commission appointed Eindhoven as candidate E-City in The Netherlands. The city's proposal was particularly praised for its impressive number of pilot projects, the quality of public-private partnerships, the balanced composition of the area and the sound organizational concept. One year later, in the summer of 2001, the region was officially appointed an e-city experimental zone. The national government decided to support the project with €45.5m in the following five years.

In February 2002, the original project organization – e-city Foundation – was transformed into a limited company, in which 27 public and private shareholders participate. Each shareholder has brought in €45,400. The project will run for five years.

Aims of the Project

In its proposal, Eindhoven defined the principal aim of the e-city to arrive at 'better living, better working and better learning' by deploying information technologies. Furthermore, the project should create conditions for accelerated time-to-market of new services, the testing of innovative ICT applications in the field of consumer services and infrastructure. The e-city's core area has a population of 84,000 people (38,000 households) and covers parts of Eindhoven and Helmond. It includes residential areas with very different characters, two city centres, business parks and office locations, as well as the campus of the Eindhoven University of Technology, and the polytechnic. It contains new subdivisions, but also reconstruction areas. The diversity may reveal differences in the way people deal with new technologies and new services, and may yield important lessons for companies and government. The size of the area is such that a critical mass of users with broadband access is reached: this is a very important condition to serve as a test-bed for new services. The e-city project has two main components: infrastructure/access provision and service development. Regarding infrastructure, the aim of the project is to deliver broadband (defined as 2Mbps minimum) to the 38,000 households in the area.

'Access' implied that each inhabitant of the e-city would have access to digital services by computer, television or mobile phone. Access policies would be supplemented by all kinds of training and support initiatives, to improve the digital literacy of the local population.

Content development was an integral part of the scheme. According to the strategy, the area should become a place where new services that add to the well-being of the population are developed, tested and implemented. The programme focuses on accelerating and improving service provision in the fields of services, health care, welfare and education/training. Where possible, use is made of already existing e-services or initiatives.

3 Governing Content

Introduction

In this section, we will describe and analyse the City of Eindhoven's policies regarding electronic content. We make a distinction between different types of content policy. The first concerns the development of online content by the municipality itself. This includes digital services for citizens, but also e-democracy initiatives. This type is often referred to as e-government in the strict sense of the word. Secondly, we review Eindhoven's efforts to promote the production of broadband content (as part of the e-city project). Next we focus on local efforts to promote the formation of online communities. Finally, we will describe Eindhoven's role in the organization of content.

Government Services Online

E-government – bringing municipal digital services online and promoting e-democracy – is one of the priorities in Eindhoven. On a national level, Eindhoven is one of the leading cities in this respect. Together with three other cities, the city was appointed by the national government as 'superpilot' city. This implies that Eindhoven receives national funding to speed up its e-government strategy. In this section, we will further comment on the local e-government initiatives.

National e-government policies
The Dutch government has the information society development high on its agenda. Within this broad theme, e-government is a key subject which receives generous attention. The coordination of e-government initiatives is in the hands of the Ministry of the Interior. This ministry has set up a dedicated ICT-implementation organization for the public sector, ICTU. Many programmes run under this umbrella. A key programme deals with electronic identification solutions and electronic security issues. At the time of writing the 'public key infrastructure' (PKI) is being tested in a laboratory situation. The first pilot projects were to be launched by the end of 2002 (PKI, 2001), with digital identity being integrated into a smart card that enables transactions with public agencies. One of the reasons for the development of this PKI was to have a single, reliable identification infrastructure and to avoid the development of many different identification schemes. However, several public organizations did not want to wait for this national security system and had already introduced other types of identification and signature. An example is the Dutch Tax Service, which already allowed the transmission of sensitive tax information through the Internet using a digital signature system.

A second key national project is Public Counter 2000 (OL2000). It aims to establish demand-oriented 'one-stop shops' at all kinds of government organizations, to stop fragmentation of the services on offer to citizens. It also plays a role in helping cities to introduce online services. Among many other things, a list of online municipal products and services (a catalogue) has been drawn up. The organization has developed practical instruments to set up online services and help public agencies that want to introduce them. The website also contains a list of companies with expertise to implement e-government solutions.

Third, national government had put a high priority on the dissemination of good practices of e-government. A centre of expertise was created as a platform for knowledge and experience exchange on innovative decision-making. Also in the OL2000 project much attention is paid to the practical implementation of e-government solutions. On the Internet, all kinds of user guides and good practices are downloadable, for instance on 'how to create a single counter', or 'how to develop proactive services' (see www.ol2000.nl).

A fourth relevant action by national government is the appointment of three 'superpilot regions'. These areas should build a lead in electronic services delivery, and serve as examples for other cities and towns in the country. The three superpilot cities are The Hague, Eindhoven/Helmond and Enschede, each of which obtained a subsidy of €2.7m.

E-government in Eindhoven

Eindhoven has to develop a 'digital counter' where citizens can submit and retrieve information. The ambition is to offer a total of 293 services/products that are listed in the OL2000 municipal product catalogue. The city has adopted a 'growth model' to introduce the services. It will start with services for which identification is not necessary, without digitalizing back office operations. In the longer run, more 'difficult' services will be introduced and the back office will be integrated with front office operations. Often, information in the back office is not complete, or not digitalized. This makes integration of back and front office impossible in the short run. For some services, a proactive approach will be taken. Citizens will receive a message (by e-mail or SMS) when travel documents or driving licenses need to be renewed. In the longer run, the municipality will introduce customer relationship management systems. Some expect that the proactive approach will lead to increased demand for municipal services. On the political level, there is strong commitment for speeding up e-government. However, in the communication and cooperation between departments, things go less smoothly. By the end of 2000, the city presented an action list of 22 points that formed the 'public' input of the e-city project. Many of them, however, have been seriously delayed. The proactive e-mail/SMS service had been postponed; citizens would be able to check the status of their construction permit, but this is still impossible (*Eindhovens Dagblad*, 6 December 2001).

Promoting the Development of Broadband Services

In Eindhoven's vision, ICT can contribute to the quality of life in the city in many respects. From this perspective, it supports ICT service development in a number of ways. In this policy, the focus of the city is to promote the development of broadband content, as part of the 'e-city' project. Companies (or other organizations) in the region that develop broadband content can get subsidy of 40 per cent of their R&D costs, at a maximum of €100,000. In total, €22.7m is available for broadband service development. By the summer of 2002, 50 applications for subsidy had been submitted (Emerce, 2002). By actively supporting the development of broadband applications, the city hopes to build a lead in this future growth market.

An example of a project submitted for subsidy is called 'Vlinderflats Internet TV'. Local volunteers wanted to make a weekly TV programme to be broadcast on the Internet. The idea behind it is to stimulate social cohesion in the area.

The 'broadband demonstration centre' plays an important role in the city's policy to promote the development and use of broadband applications. In this centre, visitors or groups can see broadband applications in action on a number of computer terminals. The centre is meant to show the added value of broadband, but it is also a platform for software developers to demonstrate their latest applications.

Local Virtual Communities

In the e-city initiative, the formation of local digital communities is one of the objectives. It is hoped that inhabitants of the e-city will start to interact with each other on the Internet when they have an 'always on' broadband connection. This would lead to new forms of social interaction and contribute to the social cohesion in the city. Private companies are interested in participating in these local virtual communities. The Rabobank – one of the largest Dutch banks – regards the e-city as a very interesting test-bed. The company views cyberspace as a domain where local companies and citizens interact in new ways. The e-city, with its envisioned 84,000 inhabitants always online with broadband, could yield important lessons for the bank. It could show how interactions and transactions among citizens and companies may change in the future. For instance, banking services could become an integrated part of local e-commerce.

For the Rabobank, as well as for many other firms, Eindhoven's e-city is interesting because it promises a unique mass of always-online broadband users. It is the 'always-online' feature which is key for the bank, because this will facilitate virtual communication. However, as we will see later, one of the key problems in the e-city is the slow progress of citizens' broadband access.

One of the local housing corporations contributes to virtual local community formation as well. It has launched a website with links to all kinds of facilities and amenities for citizens on a neighbourhood basis. Local community organizations are also invited to be present on the website. In addition, the site is a platform for discussions amongst citizens. The housing corporation developed the platform as a set of independent modules and plans to sell it to other corporations as well. The housing corporation has an interest in investing in this: in its view, it increases social cohesion and citizens' commitment to the neighbourhood. This improves the social atmosphere of the neighbourhood and ultimately, the attractiveness and value of the houses.

Content Organization: Eindhoven on the Web

How is Eindhoven represented on the web, which actors are involved in its web strategies and how is the 'local content' organized? In this section, we will discuss the city's 'web content management' as far as the City Council is involved.

The city's main Internet site is located at www.eindhoven.nl. This site is developed and maintained by the municipality: it primarily offers information on municipal developments and projects. Transaction features include online forms. No English information is available. It contains an electronic newsletter and allows for leaflet ordering. A number of forms are available in electronic form, for instance for building permits and rent subsidy applications. On the site, citizens can calculate local property tax or report moving to a new address. The site contains all kinds of information and forms, applications for travel documents, and some other services. The website is mostly in Dutch, (real-time) information on public transport is missing, there are no city maps and the overall website structure is still focused on the supply side of information and services. On a separate website, the city publishes all kinds of official municipal documents, for instance, council decisions, notes of council meetings, etc.

The city's tourist website is www.vvv.eindhoven.nl. The site is managed by the Tourist Board. It allows for hotel reservations and contains an online shop where browsers can buy brochures, maps and souvenirs. For the film, theatre and arts agendas, the relevant website is www.uitineindhoven.nl. The site contains an excellent search engine, but provides hardly any interactive features. Online booking is not possible on this website. For each event, a link is provided to the institutes.

Www.kenniswijk.nl is the website for the e-city. It offers news on the developments of the projects as well as online application forms for subsidies and a chatroom.

4   Governing Access

Citizens of Eindhoven should be prepared for the information society. This is one of the ambitions of the city. It can also be found in the e-city initiative. E-literacy among the population is encouraged by offering citizens a maximum subsidy of €68 when they take a computer or Internet course.[1] It is up to the citizen to decide

---

[1]   Of this amount, €45 is unconditional; €23 is added when the participant graduates.

where the course is followed and on what level. By November 2001, subsidies were given to 300 people (source: interview).

Several organizations offer courses. One of them is SeniorWeb, a voluntary organization that helps elderly people to use computers and Internet. One volunteer is available for every four participants. They operate throughout the country and have 1,000 volunteers in total. The Eindhoven branch is one of the largest. In 1999, SeniorWeb was also involved in the establishment of Internet cafés in homes for the elderly. Recently, SeniorWeb started an initiative to offer computer courses at home for people who are less mobile. A key aim is to reduce loneliness among elderly, handicapped people and some immigrant groups. The City of Eindhoven covers 50 per cent of the costs.

Another initiative is the digitolk. This project connects local community centres to the Internet. The centres offer web and mail services and beginners courses for Internet and computers. School children can do their homework on the computers in the centres (www.digitolk.nl).

5 Governing Infrastructure

Infrastructure policy is the cornerstone of Eindhoven's e-city. As mentioned, the e-city's key ambition was to connect 84,000 people (38,000 households) to broadband. Broadband was defined as 2Mbps minimum. The strategy consists of three parts. The first part concerns the promotion of demand by offering discounts for broadband subscribers in the e-city area: €13.6m is available for this. Second, demand is promoted by subsidising the development of local broadband applications and services. Third, it is tried to encourage the supply of fibre-optic networks to the home by fixed, mobile and cable operators. The latter category proved the most difficult.

Before the launch of the project, in spring 2000, several organizations showed interest and signed up as partners in the e-city project proposal. However, after Eindhoven won the project it proved difficult to motivate the partners. For one thing, two of the telecom partners – UPC and KPN – had serious financial difficulties and became more reluctant to invest in broadband infrastructure. Another partner, the Swedish Bredband, decided to withdraw from The Netherlands and concentrate its efforts on the Swedish home market. UPC, the local cable operator, does not seem to have much interest in investing in fibre-optic networks to the home: it would cannibalize the cable Internet market. At the time of writing, KPN is unable to invest given its precarious financial position. In sum, by the end of 2001 the prospects of Eindhoven's broadband strategy looked grim. Citizens grew impatient, because none of the promises seemed to have been fulfilled.

A new managing director for the e-city project was appointed, whose ambition was to achieve concrete results. He managed to bring companies to the table in order to at least realize some of the original Kenniswijk ambitions.

In spring 2002, finally, a small pilot project was started and 360 housing units of two housing corporations in de Vlinderbuurt area were connected to 10 Mbps (up- and downstream) broadband. The infrastructure of Bredband (which was still in place) serves as basis.[2] Users pay a monthly fee of €25, for which they get e-mail, homepage space, other services and help-desk support. Key partners are the housing corporations (they connect the homes), Volker Stevin Telecom BV (which created a broadband link between the homes and the Technical University), and Via Networks (will operate as the ISP). The e-city organization is responsible for the organization and promotion of the development of broadband experiments.

In the summer of 2002 a consortium of companies[3] announced the connection of another 1,000–2,000 dwellings to fibre-optic networks, with a 10 Mbps connection (up- and downstream). Individual households in the area can apply for a subsidy of €800. Of this amount, €500 is meant to subsidise the physical construction of the fibre-optic link. This will not cover the total cost of the connection. The big question will be how many households will apply. If the number of applicants is large enough, the project will be extended to 15,000 households in the Kenniswijk area.

Some other actors are active in providing fibre-optic connections to the home. One example is a housing corporation that connected 250 houses to fibre. Also the 'close the gap' initiative enables house owners in a block to buy the fibre-optic network for around €800 each.

6 Conclusions

Eindhoven's e-strategy is comprehensive and ambitious, and is perfectly in line with the city's slogan 'leading in technology'. The city was one of the first in Europe to realize the potential of broadband and to create a proactive policy to promote its construction and use. The approach of working on broadband infrastructure, access and content simultaneously has a lot of potential.

This case study shows that relatively little progress has been made. By the time of writing, the city had not managed to create a critical mass of broadband users in the city and the number of newly-developed broadband applications is discouraging.

This can partly be explained by a dramatic deterioration of market conditions by the second half of 2000. The dotcom crash, the debt burdens of operators (partly due to UMTS auctions) and a general recession slowed the willingness of the telecom partners to invest.

---

[2] When Bredband was still active in The Netherlands, the company already had 50 subscribers in these properties.

[3] KPN (the Dutch telecom incumbent), KVWS (a housing corporation), and BAM-NVM (construction company).

The framework of analysis states that leadership is important in realizing ICT projects. This proves true in the Eindhoven case. The new, strong project manager has managed to restart the project and achieve some concrete results. However, they are a far cry from the initial ambitions of the project. There is a risk that the public at large will turn its back on the project, which would further complicate the realization of the ambitions.

The quality and effectiveness of public-private partnerships is another success factor in realizing large ICT projects. This proved to be the case in Eindhoven. In the design stage of the project, the public sector (the project bureau) took the lead in involving partners, many of them private companies, and wrote a 'bidbook' that in the end won the national competition for the money. After this stage, the project had to be elaborated into concrete actions. This stage took one full year and was not particularly successful. It proved that the commitments of the various partners were weak. Almost every telecom market player was involved, but was there also the will to really invest? We have the impression that many actors joined the project not because they wanted to invest substantially, but rather to 'wait and see what happens'. Some of the key partners, notably the telecom companies, had entered troubled waters and were less likely to invest. The e-city organization did not manage to break the deadlock and much time passed. Meanwhile, inhabitants of the e-city grew either impatient or disinterested. During this stage, there seemed to be a difference in speed and time horizon between public and private sectors. Whereas private sector participants are interested in concrete, feasible projects on the short term, the public partners have a longer time horizon and tend to think in terms of more abstract visions or strategies.

Representatives from local communities and citizens' organizations complained about the bad communication of the e-city project management. The long time lag between the presentation of the e-city plans and the realization of the projects in particular has worried many people. The project raised expectations that could not be fulfilled. Furthermore, officials from the e-city project have not involved existing organizations who were active in helping people to get online.

The Kenniswijk project thus shows that having private partners in the project is not a sufficient condition: what also matters is their commitment to really contribute. In the next stage of the project, partners were forced to show commitment and buy a share in the organization and were much more motivated to produce concrete and feasible projects. This business-like approach has speeded up progress.

Eindhoven wants to be leading in technology, but our case study shows that the sophistication of the municipal e-services does not always meet that standard. We found that the city's own use of technology in e-government is reasonable from a European perspective. However, concerning broadband there is a large gap between the city's 'leadership' ambitions and its own use of broadband in e-government. The number of electronic local government services is limited and the sophistication of online tourist services and semi-public services leaves much to desire. This undermines the credibility of the city's policies. Stronger efforts could improve this situation. 'Leadership by example' could raise the profile of Eindhoven as a forward-looking and progressive city and also positively encourage the use of broadband by the local population.

For the future, prospects are good, as the key ingredients of 'organizing capacity' seem to be in place. The city is ambitious, there is strong leadership (both on the project level and the municipal level) and the networks with the private sector and the knowledge institutes and other stakeholders (like housing corporations) in the region are well developed. These are the necessary seeds for innovative policy.

References

Adang, A.J.V.M. and van Oorschot, J.M.P. (1996), *Regio in bedrijf, Hoofdlijnen van industriële ontwikkeling en zakelijke dienstverlening in Zuidoost-Brabant*.

Emerce (2002), 'Consortium sluit 1000 Kenniswijk-huishoudens aan op breedband', 9 July, www.emerce.nl.

Het Financieele Dagblad (2001), 'Geschoolde blanken Zuid Afrika wijken uit naar buitenland', 9 July.

Kusters, A. and Minne, B. (1992), *Technologie, marktstructuur en internationalisatie: de ontwikkeling van de industrie*, Centraal Planbureau, Den Haag.

PKI (2001), Newsletter, September: www.pkioverheid.nl/informatie/experimenten.htm.

Interview partners

O.M. Aelbers, Coordinator Research and Development, Woningstichting Hertog Hendrik van Lotharingen.

P.H.C.M. van Gastel, Senior Advisor, Department of Public Affairs, City of Eindhoven.

Elies Lemkes, Projectbureau Kenniswijk Regio Eindhoven.

T. van Lier, Projectbureau Kenniswijk Regio Eindhoven.

M.P.H. Thurlings, Project Manager e-HRM, Rabobank.

L. van Tongeren, Secretary, SeniorWeb Eindhoven.

Ton Veth, Center for Electronic Business Research and Application, Technical University of Eindhoven.

G.J.C Vos, Senior Advisor, Project Leader, E-government, City of Eindhoven.

Chapter 6

The Case of Johannesburg

This chapter first sketches the demographic, social and economic conditions in the city of Johannesburg and lists the main challenges the city faces. We also describe the national and regional context within which the city operates. This helps to understand the context in which the city pursues its ICT strategies. Next, we describe the city's ICT vision and analyse the e-government practice towards content, access and infrastructure. Finally, we draw some conclusions and policy recommendations.

1 Introduction

Johannesburg is situated in the centre of the Gauteng province, only 50km away from Thswane Metropol (Pretoria), the administrative capital of the Republic of South Africa. Around 80 per cent of the foreign visitors to South Africa arrive via Johannesburg International Airport. However, due to the capital's reputation for crime and poverty, tourists tend not to spend much time there.

During the Apartheid era, Johannesburg was divided into seven white authorities and four black townships. There was a sharp contrast between these local authorities in racial, economic and social senses. Apartheid was abandoned not very long ago, in 1995. It left deep traces in the economic, social and administrative structures of the country. In Johannesburg City Council, however, in 2002 it was no longer 'fashionable' to blame the urban problems on Apartheid. People preferred to look ahead, an attitude which is clearly expressed in the strategic vision and government practice in Johannesburg. But it cannot be denied that there is a legacy.

In 1995 the first post-Apartheid, ANC-run City Councils were established. The 11 local authorities were joined into one central metropolitan City Council and four decentralized, district City Councils. The municipal boundaries were expanded to include wealthy satellite towns like Sandton and Randburg, as well as many townships such as in Soweto and Alexandra and informal settlements like Orange Farm. The idea was to create a single metropolitan tax base. Revenues from the wealthy, traditionally white areas would become available to support the services needed in the poorer black areas. However, although the city districts were a subdivision of the metropolitan area, the district City Councils enjoyed substantial autonomy, and the principle of 'One City, One Tax Base' did not work.

This local government system was part of the new constitutional system of South Africa, designed at the Kempton Park conference. It required the representation of all parties on all Council committees. This structure quickly ran into trouble. Powers could only be delegated to committees, not to individual councillors. Accountability and transparency were lacking, inexperienced management was overambitious and spending was too high. Furthermore, services were duplicated in the city districts. The biggest problem, however, was collecting taxes and revenues for public services. Over the years a 'boycott' culture had rooted deep in society, not only in the poorer districts but just as much in the wealthy areas. The loss of revenues for the city has been estimated at more than 1bn Rand per year.

In 2000 the national government decided that there were too many local councils duplicating too many functions. Councils around the country were amalgamated, reducing the number of local authorities from 843 to 284. In this process, the Municipal Demarcation Board created 'unicities' in the metropolitan areas of Cape Town, Durban, Pretoria, the East Rand and Johannesburg.

The Municipal Organization

The unicity constitutes the metropolitan area of Johannesburg as we know it now (see Figure 6.1). All municipal legislative and executive powers are vested in the metropolitan council. The metropolitan council has decentralized operational powers and functions into 11 city districts (regions). Each city region has a People's Centre and is operationally responsible for the delivery of health, housing, sports and recreation, libraries, social development and other local community-based services.

According to South African legislation[1] there is a choice of two types of executive systems for metropolitan areas:

* the mayoral executive system where legislative and executive authority is vested in the mayor;
* the collective executive committee where these powers are vested in the executive committee.

Johannesburg has implemented a mayoral executive system (see Figure 6.2). Amos Masondo was elected first executive mayor in December 2000. The mayor has full executive powers, takes overall strategic and political responsibility and leads the Mayoral Committee, which functions like a local cabinet, with individual members responsible for different aspects of municipal government. Each member of the Mayoral Committee chairs a portfolio committee made up of councillors drawn from all political parties.

---

[1] *Local Government Act*, 2000.

Figure 6.1   Johannesburg metropolitan area city districts (regions)

Source: City of Johannesburg Annual Report (2002a).

Utilities
Pikitup
Johannesburg Water
City Power

Central Departments
Johanessburg Metropolitan Police Department
Emergency Management Services
Arts, Culture and Heritage
Housing
Health
Social Development

Agencies
Johannesburg Roads Agency
City Park
Johannesburg Development Agency

Corporatised entities
Johannesburg Zoo
Metrobus
Johannesburg Property Company
Fresh produce market
Civic theatre
Metro Trading Company

11 regions
Health
Social Services
Housing
Libraries
Sport and Recreation

Figure 6.2   Municipal structure

Source: City of Johannesburg (2002a).

The city manager is the link between the political and the administrative part of the metropolitan organization. The city manager's executive team consists of the executive directors for planning, community development, finance, corporate services, operations and contract management. The city management team implements council decisions and monitors the utilities, agencies and 'corporatized' (privatized) entities of municipality.

The political leaders of the Mayoral Committee and senior management of the city manager's team meet in the mayoral subcommittees, which are organized on a portfolio basis.

Population

The city of Johannesburg Metropolitan Municipality has 2.8 million inhabitants. 40 per cent of the population is younger than 24 years old. Africans account for 70 per cent of the population; whites account for less than 20 per cent (see Figure 6.3).

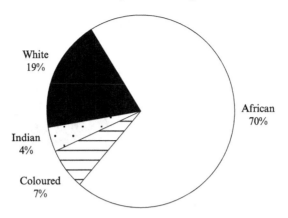

Figure 6.3   Johannesburg population (n = 2.8 million)

Source: City of Johannesburg (2002c).

Like most South African cities, Johannesburg is a divided city. The poor are predominantly black and live largely in the south or on the peripheries of the far north. More than 1 million people live in the township of Soweto, but the area contributes only 2 per cent to the municipal taxes and revenues. The middle and upper classes are predominantly white and live largely in the suburbs of the centre and the north. Some affluent suburbs offer a standard of living comparable to major cities abroad.[2]

Figure 6.4 shows the labour force skills distribution. Although the Johannesburg workforce is about 20 per cent more skilled (in terms of literacy and numeric skills) than the South African average, the overall level is regarded as far too low by the Johannesburg City Council.[3] Moreover, 76 per cent of employers in Johannesburg reported a lack of appropriately skilled staff (City of Johannesburg, 2002c).

Economy

The economy of Johannesburg, like the South African economy, used to be dominated by mining (gold, diamonds) and agriculture. Agriculture and mining continue to be important providers of employment, but over the last decade services have become the most important contributors to GDP. The services economy ranges from an advanced financial sector to a developing tourism sector, which has significant employment potential (*The Economist*, 2001). Since the discovery of gold at the end of the nineteenth century, Johannesburg

---

[2]   Population data are based on statistics available on www.joburg.org.za/unicity/.

[3]   The OECD lists the South African educational system among the worst in the world. (World Competitiveness Report 2000.)

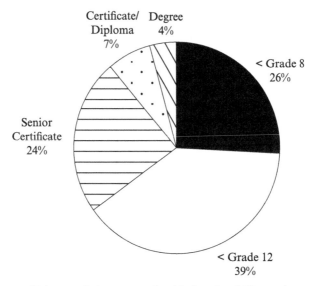

Certificate/ Degree
Diploma   4%
7%

< Grade 8
26%

Senior
Certificate
24%

< Grade 12
39%

(Below grade 8 corresponds with functional illiteracy)

Figure 6.4   Educational level of the labour force in Johannesburg

Source: City of Johannesburg (2002c).

is the commercial capital of South Africa. Its nickname is *Egoli*, which means 'place of gold'. 40 per cent of the world's gold has been found here. Nowadays, Johannesburg generates approximately 117bn Rand of GRP, which is about 40 per cent of the province of Gauteng's GRP and about 16 per cent of South Africa's GDP. Over the last 10 years the average annual GRP growth in Johannesburg was 2 per cent, which is only slightly above the national average GDP growth of 1.8 per cent. In 2000 the city provided 12 per cent of the national employment, with some 840,000 jobs in 290,000 formal sector enterprises.

The city's economy is dominated by three service sectors: financial and business services, trade (retail and wholesale), and community and social services. There is also a large manufacturing sector. Table 6.1 reveals the economic sectors' contribution to Johannesburg's GRP. Figure 6.5 shows the employment by sector.

Black Economic Empowerment

The distribution of income in South Africa is one of the most unequal in the world. In Johannesburg approximately 20 per cent of the population live in poverty, earning less than R25,000 per annum, living in shack settlements that lack proper roads, electricity or direct municipal services. Another 40 per cent live

Table 6.1   Johannesburg economic sectors

Economic sectors'contribution to GRP	%
Financial and business services	31
Trade	21
Community and social services	13
Manufacturing	15
Transport and communications	8
Public administration	2
Construction	4
Others	6

Source: City of Johannesburg (2002c).

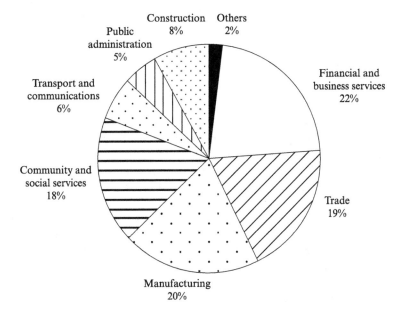

Figure 6.5   Johannesburg employment by sector

Source: City of Johannesburg (2002c).

in inadequate housing, with insufficient municipal services. Figure 6.6 shows the income distribution (of the officially employed only) in Johannesburg.

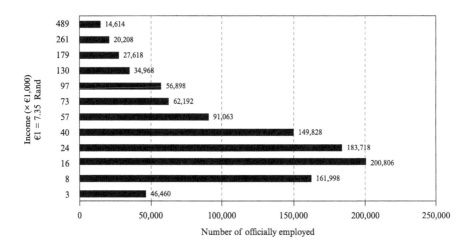

Figure 6.6   Income distribution in Johannesburg

Source: City of Johannesburg (2002c).

While economic growth in Johannesburg is higher than the national average, unemployment has risen from 27 per cent to 29 per cent over the past three years (1998–2002). Johannesburg had 393,000 unemployed in 2002.

Economic empowerment of the previously disadvantaged majority is a central theme of national government policy. Since 1994 black owned businesses have increased their share of the economy. However, at the end of the century black ownership and the number of directorate positions held by blacks declined for several reasons. As of 2001 a new, more sustainable kind of empowerment seems to be taking place. Rather than top down buy-outs and take-overs, smaller companies are being established, in the field of ICT also.

In September 2001 the Black Economic Empowerment Committee called for greater government involvement, including legislation that would set guidelines and targets. The committee pointed out that government should continue to use existing tools – such as public tenders, licences, regulations, privatizations and contracts – to stimulate the emergence of such companies. It is considered a hallmark for business transformation in South Africa. Johannesburg has already used this model when it outsourced its ICT services in 2000 (see section 4).

2 Johannesburg's Municipal Policy for the Twenty-first Century

From 1995 to 2000 the municipality of Johannesburg consisted of five separate administrations: one central and four decentralized councils. The first half of this period was characterized as a stage of *growth without sustainability*. The five more-or-less independent authorities in the region built their separate organizations and urban development plans. This has led to a severe financial crisis in the last half of the period, which started a process of trying to avoid a total collapse of the metropolitan area. The focus was on immediate actions and direct effects, with little attention on structural change: *sustainability without growth*. From 1999 onwards the local politicians and executives started looking for solutions that would overcome the fundamental problems of the city on a sustainable basis: *growth with sustainability*.

The First Strategic Plan: iGoli 2002

In 1999, Ketso Gordhan, the first city manager of the unicity of Johannesburg, started a process of integrated metropolitan development planning, which resulted in the iGoli 2002 plan.[4] iGoli 2002 was essentially a three-year strategic plan to restore the city's financial health and economic sustainability by streamlining the municipal administration. The plan provided the foundation for professionally run utilities, agencies and corporatized units. It created a distinction between the governing structures (e.g. Mayoral Committee and core administration) and operational units at the regional and central departmental level. Furthermore, iGoli 2002 introduced the outsourcing of non-core assets as well as performance management contracts for municipal agencies, utilities and 'corporatized entities'.[5]

The aim of iGoli 2002 was to ensure cost-effective service delivery by reducing fragmentation, eliminating duplication, improving accountability, managing human resources and providing performance incentives. The plan actually redirected municipal resources to under-serviced areas in the Johannesburg metropolitan area.

The Second Strategic Plan: iGoli 2010

The 10-year iGoli 2010 framework extended the internal administrative performance indicators of iGoli 2002 to external metropolitan performance indicators. The overall ambition is to develop Johannesburg into an African world-class city by 2010, operating at the level of the highest norms and quality standards

---

[4] 'iGoli' is an alliteration *Egoli* (place of gold), Johannesburg's nickname since the discovery of gold.

[5] A 'corporatized entity' is effectively a privatized company, with the city as the single shareholder.

in the world and running a competitive, internationally-oriented economy. As an 'African city' it wants to use its geography and its citizenship as a distinctive asset and a competitive advantage. Johannesburg is to be promoted as a place to live, invest and visit, thus creating jobs, development and housing opportunities.

The iGoli 2010 framework aimed at a growth of Johannesburg's economy of at least 5 per cent per year (it has been 2 per cent on average over the last 10 years). Employment should grow by at least 4 per cent each year. The iGoli 2010 framework established similar targets for each sector (health, demographics, transportation, etc), but it did not make any recommendations regarding strategic options and choices.

The Third Strategic Plan: Joburg 2030

In 2030 Johannesburg will be a *world class city* with service deliverables and *efficiencies* which meet *world best practice*.

Its economy and labour force will specialize in the service sector and will be strongly *outward orientated* such that the City economy operates on a *global scale*.

The strong *economic growth* resultant from this competitive economic behaviour will drive up City tax revenues, private sector *profits* and individual disposable *income* levels such that the *standard of living* and *quality of life* of the City's inhabitants will increase in a *sustainable* manner. (City of Johannesburg (2002c): original text in italics)

The iGoli 2010 framework was used as the basis for the development of a long-term vision and strategy for the city: *Joburg 2030*. The front page of the plan states the main ambitions of Joburg 2030, above

The *Joburg 2030* document contains an overview of the demographic and economic situation of Johannesburg, followed by a thorough analysis from the perspective of economics of urbanization and economics of localization, which builds the core of the vision. The plan outlines a strategy to implement the vision, including the required actions, programmes and interventions. A new unit – the Corporate Planning Unit (CPU) – has been set up by the City Council. It is responsible for driving the implementation of Joburg 2030.

At the time of writing Johannesburg is failing to meet its full economic potential. A major reason for this is the fact that Johannesburg was historically geared to service the needs of heavy, primary production sectors, particularly the gold mining and iron and steel industries. As these industries declined, the existing economies of localization present in the city become increasingly outdated. According to Joburg 2030, Johannesburg's economic landscape will no longer be dominated by mining and manufacturing, but by the service sector. In order to achieve this, the city must grow into an export-oriented hub, closely integrated into the global economy, with the emphasis on trade, transport, financial and business services, information and communication technology and business tourism. Two of the main challenges in growing the services sector will be the ability to generate sufficient and appropriate skills within the labour force – thus closing the skills gap – and to *upgrade information and telecommunication systems.*

To close the 'skills mismatch' between industry needs and labour force supply, the City Council aims to harness all the city's educational resources – community centres, libraries, museums – also for after-school learner support. It will also investigate ways of *increasing computer access* at all the city's schools.

In order to accelerate growth within the city the Council will take several other initiatives. First it is proposed that *a dynamic, robust and reliable information system* be developed for both private and public sector use. The current lack of reliable city-based data makes informed planning virtually impossible. A database of key economic variables and trends would contribute to better decision-making and growth across the city. Another initiative is the creation of task teams that will support the main priorities of the economic development plan by dealing with issues of *access to information, infrastructural constraints and marketing.*

3   National Context for ICT at Local Level

National governments are important players in the ICT policy fields. In the first place, they are responsible for the overall conditions in which the transition to the digital economy can take place. They set standards and legislation for digital identification, define property rights and set telecom legislation. Second, national governments can promote the creation and use of ICT throughout society in various ways. Third, national governments are important ICT-users themselves.

In South Africa, more and more government functions are being devolved from the national level to management at local level. This is partly due to fiscal and budgetary constraints at the national and provincial levels, and often leads to the delegation of 'unfunded mandates'.

The government strongly supports and encourages the country's transition towards a knowledge economy. It realizes that the deployment of information and communications technologies is an important enabler, and it is implementing ambitious plans to introduce e-government into its own organizations. A number of national government departments in South Africa have issued policy documents which elaborate strategies for the adoption of ICT in society. One of them is the e-commerce Green Paper published by the Department of Communications. Another is *Electronic Government, the Digital Future IT Policy Framework* published by the Department of Public Service and Administration (http://www.dpsa.gov.za). Finally we should mention the White Paper on Science and Technology by the department of Arts, Culture, Science and Technology (http://www.gov.za/whitepaper/1997/sc&tecwp.htm).

Furthermore, there is a strong pressure to improve the quality of service delivery. The national government's 'people first' White Paper strongly calls on local governments to improve their performance and become more customer-oriented. This illustrates that *access* is a pressing issue in South Africa. The national Access to Information Act promotes a society in which the people of South Africa have effective access to information, thus enabling them to more fully exercise and protect all of their rights. Other stakeholders, like communities and the private sector, are also exerting pressure for better service and information provision.

Telecommunications Infrastructure

From an international perspective South Africa's telecom infrastructure is poor. This is mainly due to the monopoly position of Telkom, the incumbent telecom operator. Prices are relatively high and quality of service is bad. Under current legislation, no company can construct its own cabling between two buildings, unless they are adjacent. In all other cases, they have to lease a Telkom line. Bandwidths higher than 2 Mb are extremely expensive.

Telkom operates the third largest underground network in the world. The network is made up of 156 million kilometres of copper cabling and 343,000 kilometres of fibre-optic cabling. The policy in major urban areas is to roll out fibre optics to the districts, with copper cable connections to the buildings. Higher-speed services cannot be accessed on the existing infrastructure. This means that, in order to envision a mass of home Internet users in the townships across the country, a new roll-out of local infrastructure would be required. However, Telkom does not have a policy in place that anticipates the provision of anything but basic telephone services to townships.

At present, the Department of Communications in South Africa is most concerned with 'teledensity', which measures the number of telephone lines per 100 inhabitants and was one of the conditions of Telkom's exclusivity on infrastructure. Meeting this requirement has been a top priority in Telkom's strategic operations. With an average teledensity of 10.05 per cent at the national level – and 47 per cent in Johannesburg – Telkom has in fact surpassed the required rate. However, it has done little to innovate and modernize the infrastructure.[6]

South Africa is presently catching up with the worldwide phenomenon of privatizing and liberalizing its telecommunications industry. The monopoly of the national telecom operator Telkom expired in 2002. At the time it was expected that at least one additional operating license would be granted. It has been suggested that more than one licence may be soon granted.

---

[6] Using teledensity as the most important measure of progress is not undisputed. *Universal access* appears to be much more relevant for economic development and social cohesion policies (DRA Development, 2000; Hill, 1999; Graham and Marvin, 1999).

According to a CIA estimate, in the year 2000 there were 1.82 million Internet users in South Africa.[7] This represents 4 per cent of the South African population. By Western European standards this is a low figure. However, growth is rapid: a more recent study by BMI-TechKnowledge has put the number of Internet users in South Africa at 2.6 million in 2001 (6 per cent of total population), with 1.6 million devices being used to access the Internet. Although there are no regional data available, PGWC (2001) estimated that the Gauteng province was a frontrunner in South Africa, together with Cape Town and The Western Cape province.

4 Johannesburg's Vision on E-Governance

Until 2002 e-government had not been a strategic priority in Johannesburg. As far as it existed, it was largely driven by the Gauteng provincial government. In Johannesburg at that time other issues required more attention. Information and communication technologies were being mentioned, but they did not make it into the top five of strategic priorities. At best, they were seen as a tool, not as an enabler of strategic options. However, although the mayor is not an IT-user himself, he is aware of the potential of IT to improve the overall efficiency of public administration. He also feels the need to prevent the digital divide from reinforcing existing cleavages in society and he knows that Johannesburg needs to catch the potential of the knowledge economy.

At the time of writing, the third strategic plan, *Joburg 2030*, imposes several challenges that are relevant to Johannesburg's e-governance strategy, such as:

- closing the skills gap and the skills mismatch;
- upgrading information and telecommunication infrastructures;
- increasing computer access and computer skills;
- developing dynamic, robust and reliable information systems;
- improving access to information, infrastructure and marketing.

The strategic vision *Joburg 2030* recognizes ICT as the country's and the city's best opportunity to increase effective and efficient service provision and a better quality of life for its citizens. Furthermore, ICT will become crucial to the way citizens in their individual households access information about general services, education, health and the job market. Last but not least – given the goals set out in the national 'People First' White Paper – it is an important tool by means of which local government will interact better with its citizens.

However, *Joburg 2030* takes a much broader view. It states that the issue of information and telecommunications technology (ICT) is crucial to Johannesburg's future. Information intensive industries contribute substantially to the city's GRP.

---

[7] http://www.cia.gov/cia/publications/factbook/index.html.

They require – and already use – high volumes of telecommunications. However, at the time of writing, telecommunications facilities in Johannesburg were rated amongst the poorest in the world in terms of costs and service quality. 89 per cent of firms interviewed stated that the costs of telecommunications present a 'major inhibitor to growth'. This is a pressing issue, because those particular industries – which include financial services, trade, transport logistics and business services, as well as the IT industry itself – are all future leaders of the city's service economy. For other sectors of the economy also, ICT is a potential tool to unlock many underperforming industries. It would also offer a significant opportunity to catapult a variety of local sectors into international competitiveness.

All in all, the role of ICT is a vital input for Johannesburg's future economic growth, social cohesion and urban governance. This is why the Johannesburg City Council has been considering which additional responsibilities and political powers it should claim in this field. It is clear, for instance, that the national policies towards liberalization of telecommunications have a direct effect on Johannesburg.

The E-Government Strategy

The Information Technology Department is responsible for the city's e-government vision and implementation. Essentially, the department works in the background, setting up and improving all IT systems for the City Council and ensuring that they run smoothly. The department's major achievement up until 2001 was to systematically strengthen the City Council's IT infrastructure as *a tool* to connect the municipal offices and to support existing workflows. The IT Department attributes its success in improving the city's infrastructure to the Council's decision to make information technology a cornerstone of its drive to become a world-class African city and to commit financial and other resources to making this possible.

In 2002 the Mayoral Committee, the city manager and the IT department considered what the next logical step would be. The question was how to use ICT as an enabler of the new forms of government and business processes that the city wished to implement. This required a level of leadership and focus that went beyond what would be provided in a technical support function: *a driving force* rather than a facilitator.

User education is a first challenge. Most staff members use only a fraction of the power of the existing information systems due to fear or lack of knowledge with respect to ICT. The metropolitan, district and departmental managers face the challenge of understanding how information technology can help their organizations work better. The IT Department advises them on how to make full use of the infrastructure, how to work together with other departments and how to avoid duplication and overlap.

A second challenge is to *use ICT as an enabler for the transformation of Johannesburg* towards a world-class African city. Information technology allows people to talk to each other and work together more effectively. The IT department should be engaged in business process redesign and customer orientation, promoting a working culture that encourages innovation, interaction and integration.

The Chief Information Officer (CIO) and Director of IT is a member of the city management team, which underpins the strategic value of e-governance. The importance of IT as a business asset was further secured in November 2002 when a Chief Operations Officer was appointed. She is supposed to be the highest administrative position next to that of the city manager. Her responsibilities include – among other things – community development, the city district offices (*people centres*), information technology and the municipal call centre (*Joburg Connect*).

Outsourcing of ICT to the Sebedisana Technology Consortium

The Chief Information Officer (CIO) is the head of the municipal Information Technology Department. She has only a handful of staff members. Operations have been outsourced to Sebedisana, a consortium of IBM (51 per cent) and the Black Economic Empowerment enterprise Masana Technologies (49 per cent). Under the outsourcing contract IBM South Africa is responsible for the transfer of IT skills to the previous municipal IT-workers, who are employed by Masana (see Figure 6.7).

The outsourcing contract lasted for five years. At the start, in October 2000, the minimal fixed value of the contract was around 500m Rand (€70m), with expectations up to 720m Rand (12m Rand per month). By mid-2002, however, the actual turnover for 2002 could be estimated at 215m Rand (Table 6.2). On the basis of Johannesburg's increasing ambitions for ICT, the market value of ICT services, and with expected good performance of Sebedisana, the overall value of the contract can easily exceed 1bn Rand by 2005. The evolution of the outsourcing contract is foreseen in the so-called 'request for services' procedures (RFS). If new services (not mentioned in the contract) or other service levels are required, e.g. for network services, an amendment is added to the contract. Eventually, the basic agreement can be revised.

Although the present city manager would in principle prefer Johannesburg to have its own IT services department, the city had no other choice than to outsource its IT services in 2000.[8] The financial position of the city was extremely poor, IT management was underperforming, there were hundreds of IT costs centres spread throughout the administration, the level of internal IT services was poor, and there was an apparent lack of skills for the delivery and use of desktop and

---

[8] At national level, the government and the unions of local government workers have concluded a National Framework Agreement, which states that public option comes first. According to this principle, private sector involvement is only an option if the public sector does not – or cannot – work well enough.

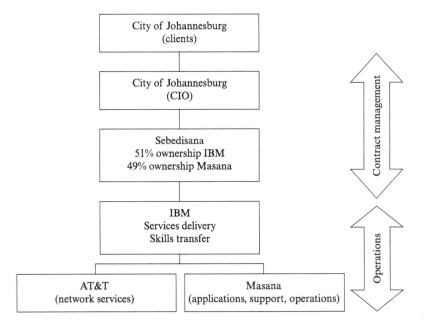

Figure 6.7   Structure of the outsourcing partnership

LAN services. Outsourcing was the quickest and the easiest way to buy in the required expertise.

However, the city retained IT development briefings for senior management and the responsibility for strategic choices. Up until 2002 the city was struggling to implement this responsibility at an appropriate level. It was quite difficult for the municipality to implement a corporate CIO unit that would address *ICT issues* from the municipal *core business* perspective, whilst being an expert in both fields – countervailing the expertise of Sebedisana. At the same time it was quite difficult for IBM South Africa to develop the business case with Johannesburg. IBM is used to address different market segments through the involvement of business partners, who actually deliver the business value to the customer. The municipal IT department was considered as the business partner for the local government segment in Johannesburg, but actually was in no position to take up this role by lack of skills, lack of mandate and lack of authority. At the same time the CIO was reluctant to allow Sebedisana to enter into direct business relations with the municipal client organizations.

The business case has probably been saved by existing personal linkages. Before the outsourcing took place, many employees of Masana used to work for the municipal departments and enterprises. They know the business and they know the people in the customer organization. In this sense Masana is a strategic partner for IBM. At the same time the relations between IBM executives and high members of the city government appeared to be very good, leading back – at least for some

Table 6.2   **Sebedisana expected turnover with Johannesburg, 2002**

Budget	Contractual basis	Rand	€
Central IT budget Departmental/ decentralized	Outsourcing contract	170,000,000	23,129,000
IT budgets	Outsourcing contract	30,000,000	4,082,000
Other budgets	Additional contracts	15,000,000	2,041,000
Total value		215,000,000	29,252,000

of them – to the time that they were united in the struggle against Apartheid. Those relations extend to the national level and are being well managed.

5   Governing Content

This section describes and analyses the creation of electronic content by the city. The main focus lies on the introduction of internal IT systems and the city's web strategy.

Internal ICT Systems

In common with many other cities in Europe and South Africa, different departments of Johannesburg's administration have their own, sometimes very large, information systems and customer databases. In consequence, multiple inconsistent versions of management information and citizens' data exist. In Johannesburg, however, the situation is much more complicated due to the fact that the metro government, which was only created in 2000, is a melting pot of five previously independent – and in fact competing – administrations. This has left the Johannesburg City Council lacking standards, information architecture and information security, as well as having a low level of IT literacy in the user communities.

Bad information systems are the cause of substantial underutilization of the *local tax base* in Johannesburg, which constitutes 96 per cent of the municipal budget (including fees and revenues). The assessed value of real estate property in the city hardly changes in the municipal books. It increases by 0.5 per cent per year, while the real economic value increases much more. In consequence, the revenues are too low – in particular for the booming parts of the city, where real estate value tends to increase most.[9]

---

[9]   Unlike Europe, annual property taxes do not exist in the Republic of South Africa. There is only a value-based sales tax when the property is sold. However, the valuation of the property is also the basis for other taxes and public service delivery fees.

Proper administration is another problem. The City Power enterprise, for instance, delivers 6.5bn Rand of electricity to the Johannesburg metropolitan area. However, due to lack of appropriate administration, 1bn Rand (€136m) is not being collected.[10] Only part of this gap is caused by customers's *payment problems*, in particular in the townships of Alexandra and Soweto. The main problem and the primary target is to get more invoices out to the new shopping malls, residential and business areas as well, many of which have not yet been properly administered.

The IT department has started a process of harmonizing the IT infrastructure, workflows and document management. The ICT services provider of Johannesburg, Sebedisana, has chosen not to replace the existing databases by a single new database. The focus rather is on an information architecture that has the potential of integrating and improving the data that are already present in the existing databases. The Electronic Account Presentation (EAP) is an example of a system that enables residents to update their personal data – such as addresses, marital status and contact details – in a central database. The system started as an e-mail system of presenting municipal bills for utilities to the customers. As a next step citizens could update their meter readings online. Many citizens appeared to take an interest in getting the meter readings right. At the same time, the quality of the personal data in the municipal systems improved. The new personal data will be used as basic data for other municipal information systems, thus gradually replacing defective data.

Sebedisana is also building a business intelligence system for the city's financial department. This system will evolve towards an enterprise resource system, enabling easier and more comprehensive management information queries from the municipal data sources. Meanwhile, virtually all departments and municipal enterprises have appointed a Chief Information Officer (CIO) and the management of information has moved out of the back offices into the Office of the City Manager.

Web Strategies

The utilities, agencies and corporatized entities of Johannesburg enjoy operational freedom. The Council sets the policy for these enterprises and defines output and service levels. It seeks a balance between operational management freedom and overall strategic requirements of Johannesburg as a whole. As far as ICT services to the citizens are concerned, the city should speak with one voice.

---

[10] Ongoing research commissioned by the South African Revenue Service (SARS), reveals that the tax gap at national level can be estimated at 55–117bn Rand (€7.5–16.0bn). SARS allocates substantial resources to developing advanced technologies and link up third party data bases in order to close the gap ('Put the resources where the risk – or possible benefit – is'.)

The IT department has planned the transformation to electronic government to take place in four stages. Much of the work will be internal and some will be targeted directly to the citizens. In the first stage, the IT department has set up publishing of static information on the City Council intranet and the Internet. This allows staff to access information when providing services to the citizens. The second stage provides for one-way provision of dynamic information. Municipal employees can request specific information from the intranet databases. The services at this stage will include a human resources portal, planning information and project tracking. In the third stage two-way transactions become possible. Citizens will be able to pay their accounts online. Tracking systems will facilitate citizens' access to information. Finally, in the fourth stage, the city will have moved close to full electronic government. Fully interactive systems will be implemented. This will include e-procurement, enabling the Council to buy online.

6 Governing Access

In July 2002 the City Council discussed the city's information technology strategy. According to this strategy, by 2005 the Internet and related solutions would be the dominant means of enabling access to government information, services and processes. The public sector will install a shared information architecture that delivers online information and services through multiple channels of communication, such as personal and handheld computers, fixed and mobile telephone connections and self-service kiosks.

Johannesburg has a responsibility to enforce the Citizens' Right of Access to Information Act at the local level. It fosters a culture of transparency and accountability. The strategic vision *Joburg 2030* states that ICT will become an important tool for interaction between local government and its citizens. Furthermore, telecommunications and information technology would become crucial to the way citizens in their individual households access information about general services, education, health and the job market. This is why the City Council had the ambition to facilitate access to computers for all schoolchildren before 2005. In addition, kiosks and other access provisions are to be implemented in public places such as neighbourhood centres.

The educational institutions in Johannesburg – mainly financed and governed by the state and provincial level – are not very active in addressing the skills gaps, which are the largest in the field of ICT and the knowledge economy. However, many private organizations are taking initiatives.

In line with the national 'People First' directives, Johannesburg has set up *people's centres*, which are municipal service centres in the city districts. For those who have access to a telephone, a single number call centre, *Joburg Connect*, was established in 2001. Finally, since July 2002 the City Council's website has provided information and services for customers with access to the Internet.

There is an initiative, supported by the government, to create Internet access points in multipurpose community centres in Johannesburg's townships. It would start with about five access points per community centre. Expectations are that it will evolve to 10 units per centre, which constitutes a total of 1,000 access points across Johannesburg. At the time of writing, however, many issues have to be resolved, such as user education, instructors, maintenance and support staff, communications infrastructure and theft prevention.

7   Governing Infrastructure

Johannesburg currently has a density of 47 per cent privately-owned phones. Access to telephony through other means (such as pay phones, cell phones or friends' phones nearby) averages at 95 per cent for the city as a whole, with the lowest rate in Orange Farm at 84 per cent and the highest in Sandton at 97 per cent.

Sebedisana maintains the city administration's internal networks, creates infrastructure and supports staff members with computer use. Through the outsourcing contract the information systems of the city have become stronger, better connected, faster and more reliable. Approximately 3,500 computer users in almost 200 City Council buildings are linked through local area networks (LAN) and wide area networks (WAN). The networks and servers are being upgraded and over a three year period the desktop computers are being replaced with new ones. All staff can use e-mail and access information through the intranet. Connectivity to the Internet is underway.

At the urban level, the ambition of Johannesburg is to become a world-class city in terms of economic strength, social cohesion and governance. Reform in the telecom situation in South Africa is a precondition to realizing this ambition, as infrastructure is one of the basic provisions on which a knowledge economy thrives. The city needs increased bandwidth and better service provision at lower costs, as well as a competitive advantage in telecommunications. Johannesburg intends to do everything in its capacity to influence the relevant national policies. It seeks cooperation with other major South African cities, e.g. through the South African Cities Network (SACN), and it considers issues such as Rights of Way, coordination of utilities' activities, and building regulation amendments that take cabling needs into account.

The ambition, however, is much higher. As a first step, the Corporate Planning Unit (CPU) has developed a position paper about the role of local government in the telecom market, opening up the possibility of becoming an active participant in the sector. This position would be circulated and lobbied at provincial and national government levels. Once the city's viewpoint on this issue has gained political support, Johannesburg will work on a business plan. At that stage the CPU will look at a range of options. These include the possibility of the Council itself developing a suitable infrastructure and then leasing spare capacity to private operators (like Austin, Texas and Stokab in Stockholm), creating a public-private partnership to provide city-wide services (like Anaheim, California), entering into contract arrangements with local providers (like Boston, Massachusetts), setting up ICT clusters or villages (like Kuala Lumpur, Silicon Valley and Bangalore) or setting up the city as a CyberCity (like Toronto). The idea is to have the business plan proposition ready when a third telecom operator's licence is granted in South Africa.

8   Conclusions and Recommendations

In this chapter we have analysed Johannesburg's ICT strategy. It is clear that Johannesburg is a city confronted with considerable challenges, particularly compared to the European cities that we also analysed. One of the key challenges is to close the skills gap – the 'digital divide' – to improve the quality of life for the growing number of inhabitants, and to guide the city into the knowledge economy. Another challenge is to manage the large unified city in a comprehensive way.

The third strategic plan – *Joburg 2030* – links some key issues of ICT (increasing computer access and IT skills, upgrading the telecommunications infrastructure and developing reliable information systems) to strategic urban priorities, such as closing the skills gap, creating equal opportunities for previously disadvantaged groups and transforming Johannesburg into a first-class African city with a knowledge-based economy. As yet however, the turn-around of the back-office – implementing ICT as an enabler of renewing municipal business – remains a point of concern. In 2002 the Mayoral Committee, the city manager and the IT department considered how to use ICT as an enabler of the new forms of government and business processes that the city wished to implement. This requires a level of leadership and focus that goes beyond what would be provided in a technical support function: *a driving force* rather than a facilitator. *User education* is a first challenge.

Content and Services

Our analysis suggests that the city – at least until recently – did not have a high strategic level of thinking and acting in this field. The ICT strategy had a strong focus on process efficiencies, but it did not seem to be very well related to the overall business strategy for public services. This appears to have changed around 2002, when a four-stage electronic government strategy was put in place. It seems to be a sensible strategy, gradually developing the services of the city administration, the agencies and the corporatized enterprises towards fully interactive systems over a period of three years. Furthermore, efforts have been put in place to make sure that the municipal entities harmonize the implementation of ICT services to the citizens. Meanwhile, Sebedisana is improving the quality of the corporate databases and implementing pilot projects that fit well in the municipal strategy, such as the electronic account presentation and meter readings online.

Access

The Citizen's Right of Access to Information Act leads the way to implementing universal access at all levels of government in South Africa. In accordance with the national 'People First' directives, Johannesburg has set up people's centres across its territory. In addition, ICT is being put in place to allow access over the telephone (*Joburg Connect*) and, more recently, the Internet (www.johannesburg. gov.za). Of course, closing the skills gap and creating equal opportunities for previously disadvantaged groups are two of the main challenges. Internet access points are being created in Johannesburg's multipurpose community centres. As yet, however, many issues have to be resolved, such as user education and training of teachers. There is room for closer cooperation with educational institutions. Although they are mainly financed by the state and provincial level, they could be more active and more effective in the field of bridging the digital divide. Other issues of concern are security – including the protection of Internet access points against crime – and the development of adequate communications infrastructures.

Infrastructure

The present situation of telecommunication infrastructure and respective legislation in South Africa hampers the necessary development of high capacity internal and external communication infrastructures in the city of Johannesburg. It also interferes with the ambition of Johannesburg to become a world-class African city. Reform in the telecom situation in South Africa is a precondition to realizing these ambitions. As has been stated in *Joburg 2030*, Johannesburg should do everything in its capacity to influence national telecom policies. In doing so, it could very well cooperate with other major South African cities which have similar problems and ambitions. International comparisons and benchmarks can be of great help here.

We did not find attempts to integrate the networks for voice (telephony) and data communications that the city of Johannesburg was using at the time of our investigations. Apparently, the liberalization of telecommunications in South Africa has not yet reached a stage at which 'voice over data' is an option. The potential benefits, however, are considerable. The issue should be included in Johannesburg's lobby to influence national telecom policies. In the meantime, however, as long as improved legislation is not in place the city of Johannesburg and Sebedisana might consider implementing voice over data in partnership with an appropriate third party.

Strategic Public and Private Partnerships

In many cases the integrated services that information and communication technologies can deliver – if they were to be exploited to the full – require effective interaction between public and private organizations. The outsourcing contract for IT services – and the way it is actually working – is a good example of a strong public-private partnership. It involves the municipality, Sebedisana, IBM South Africa and Black Economic Empowerment company Masana Technologies. This partnership does not only deliver IT services to citizens and municipal workers. It is in fact a strong instrument for bridging the skills gap for the developers and the users of IT applications. It is also a potential enabler of innovation in business processes. To use this potential to the full, however, would require a stronger integration of IT management issues in management skills across the Johannesburg administration, including the utilities, agencies and corporatized entities.

It may be clear that the e-government strategy requires a clear vision about the citizens, the municipal organization and the use of technology. In the case of Johannesburg this vision is well founded in local and national strategies, such as *Joburg 2030*, *People First* and *Electronic Government – The Digital Future IT Policy Framework*. At the time of our interviews however (mid-2002), we did not find an explicit e-government strategy in Johannesburg itself. The city relied heavily on the outsourcing contract with Sebedisana. As the knowledge partner in this contract IBM South Africa was actually leading the way to e-government. It had difficulties in finding an appropriate 'business partner' in the municipal organization with whom it could develop e-government at the strategic level. The business case has probably been saved by existing personal linkages. Before the outsourcing took place, many employees of Masana Technologies used to work for the municipal departments and enterprises. At the same time, relations between IBM executives and high members of the city government appeared to be very good. These bypasses, however, can not conceal the fact that there is room for improvement on the municipal side. The central IT department could take a more strategic role – by, for example, imposing ICT as an enabler of change, defining the ICT architecture, setting the standards, stimulating innovative use of ICT, managing the outsourcing contract and enforcing the framework for departmental, business-driven, implementations – leaving appropriate responsibilities and mandates to the municipal departments and enterprises. Perhaps the appointment in October 2002 of a Chief Operations Officer (the highest administrative position after the city manager) is an improvement.

Apparently there is also room for improvement in IBM's marketing strategy towards large customer organizations in the public sector – perhaps not only in South Africa. Rather than working through business partners and outsourced IT enterprises, IBM should directly show its face – and lend its ear – to the actual users in this particular market. In the case of Johannesburg, Sebedisana and IBM South

Africa could further developed their customer relations in this way. But there may also be a case for and IBM-EMEA and IBM Government Industry worldwide.

Outsourcing of IT Services

According to the National Framework Agreement, private sector involvement in South African public services is an option if the public sector cannot do the job well enough itself. However, the public option comes first. As a consequence the issue of outsourcing should always be an explicit choice. As we have seen, in 2000 Johannesburg made such an explicit choice – on solid grounds. Outsourcing IT services was the best way of acquiring the technological expertise needed to develop responsive government and good quality public service delivery. The city had no other choice than to outsource its IT department. However, the issue will remain on the political agenda.

The issue of outsourcing should be approached with the utmost caution. For the sake of responsive government and public service delivery it is indispensable that public authorities have direct access to – and control over – the application developers, the operators and the engineers that make it work. One of the issues at stake is that sooner or later the mother company of an outsourced IT department will be inclined to impose its own standards of management, technology, information architecture and pricing. Furthermore, in a free and competitive market there is always a risk that ownership of the outsourced IT department or its mother company will change into hands beyond the control of the customer organization.[11] This is not to say that outsourcing is not an option. Rather, the issues should be properly addressed in the strategic management of the outsourcing contract, the partnership model, the ownership and the influence over the IT services enterprise.

References

City of Johannesburg (2001), *Johannesburg an African City in Change*, City of Johannesburg/Zebra Press.
City of Johannesburg (2002a), *Annual Report 2001/02*.
City of Johannesburg (2002b), *Integrated Development Plan 2001/2002*.
City of Johannesburg (2002c), *Joburg 2030 – A Vision Statement for the City of Johannesburg*.
DRA Development (2000), *Telecentres 2000 – The Way Forward*, Johannesburg.

---

[11] Some of the cities that outsourced their IT services department in the past regret they did. At least one of them (Leipzig) even bought back its IT department a few years after it had been outsourced. The city of Barcelona has chosen to give its IT services department a central position in the municipal structure. It is a safeguard for strategic use of ICT. At the same time the issue of outsourcing IT services has never disappeared from the political agenda.

The Economist (2000), 'Government and the Internet: Haves and Have-nots', 24 June.

Financial Times (2002), 'Johannesburg – A City of Contrasting Realities', 7 August.

Graham, Stephen and Marvin, Simon (1996), *Telecommunications and the City: Electronic Spaces, Urban Places*, Routledge, London.

Hill, Sir Anthony (1999), Report of the Focus Group to ITU Study Group 3 (The Hill Group), http://www.itu.int/osg/spu/intset/focus/index.html.

OECD (2000a), *Information Technology Outlook 2000*, OECD, Paris.

OECD (2000b), *World Development Report 2000*, OECD, Paris.

Provincial Government of the Western Cape (2001), 'The Cape Online Programme', version 3.9.

Republic of South Africa (1998), *The White Paper of Local Government*, 9 March.

Websites

www.communitysa.org.za – website for community ICT projects in South Africa.

www.cia.gov/cia/publications/factbook/index.html – fact sheets about South Africa.

www.dpsa.gov.za – Department of Public Service and Administration, RSA.

www.economist.com/cities – The Economist's cities guide, including Johannesburg.

www.euricur.nl – website of the European Institute for Comparative Urban Research.

www.ft.com/johannesburg2002 – website of the *Financial Times*, featuring Johannesburg.

www.gov.za – general website of South Africa government online.

www.ibm.com/za – IBM South Africa's website.

www.johannesburg.gov.za and www.joburg.org.za – official website of the city of Johannesburg.

www.sacities.net – website of the South African Cities Network.

www.salga.org.za – website of the South African Local Government Association.

Interview Partners

Andrew Boraine, Special Advisor to the Minister, Ministry of Provincial and Local Government, Cape Town.

Goodnews Cadogan, Compliance, South African Revenue Service (SARS), Pretoria.

Kgotso Chikane, Spokesperson for the Executive Mayor, Johannesburg City Council.

Rolf Dauskardt, Institute for Housing and Urban Development Studies, Rotterdam.

Anne C. Duiker, Senior Manager, Johannesburg Receiver of Revenue (SARS), Johannesburg.

Cllr A.K.L. Fihla, Member of the Mayoral Committee, Chairperson of Finance, Strategy and Economic Development, Johannesburg City Council.

Heather Fuller, Executive Director, Information Technology Department, City of Johannesburg.

Cllr Brian Hlongwa, Member of the Mayoral Committee, Chairperson of Municipal Services Entities, Johannesburg City Council.

Roland Hunter, Executive Director for Finance, City of Johannesburg.

Tim James, Project Executive for Johannesburg, IBM South Africa, Johannesburg.

Keith R. Kenneth, Manager International Relations, South African Revenue Service (SARS).

Sunita Manik, Information Technology Department, South African Revenue Service (SARS).

Zamile Mazantsana, Client Executive Public Sector Business, IBM South Africa, Johannesburg.

Johan van der Merwe, Project Executive Strategic Outsourcing, IBM South Africa, Johannesburg.

Pascal Moloi, City Manager, City of Johannesburg.

George Mxadia, IBM South Africa, Johannesburg.

Excel Shikwane, Public Sector Executive, IBM South Africa, Johannesburg.

Sandhya Sonne, IT Architect Business Innovation Services, IBM South Africa, Johannesburg.

Judy Sybisi, SALGA.

Chapter 7

The Case of Manchester

1 Introduction

In this chapter, we will describe and analyse Manchester's e-governance strategies. We will start in section 2 with a general context description of the challenges that the city of Manchester faces. This puts the city's e-governance efforts into perspective. In section 3, we summarize the city's ICT strategy. In sections 4, 5 and 6 we analyse the issues of content, access and infrastructure respectively. Section 7 concludes.

2 Manchester: Profile of the City

The city of Manchester is situated in the northwest part of England. It counts 431,052 inhabitants. The city forms the heart of the Greater Manchester region with more than 2.5 million inhabitants. It is known as the cradle of the Industrial Revolution. However, these glory days are over. At the dawn of the twenty-first century, the city is struggling with major problems associated with the economic restructuring that characterized the last quarter of the twentieth century. At the same time, the city is trying hard to revitalize and regenerate its economy.

Spatial-economic Structure

The economic structure of the city is dominated by services. Growth sectors of the last few years have been the cultural industries, tourism and the ICT sector (Manchester City Council, 2001e). The city region is popular with e-commerce companies. Tourism earns Greater Manchester some €2.4bn annually, accounting for 6 per cent of the city region's GDP, and an estimated 45,000 jobs (Manchester City Council 2001ed, p. 9). Most visitors come only for the day, for football, shopping or a concert. Despite relatively high growth rates in these sectors, the economic performance of the city is still poor from a national perspective. With 7.7 per cent, the unemployment rate is double both the regional and the national average (Manchester City Council, 2001a). In some wards however, it reaches nearly 20 per cent. Exports per capita fall short of the UK average. The average weekly income of people in work is €300; well below the UK average €480 (Manchester City Council, 2001a).

In spatial terms, Manchester is a divided city, with sharp contrasts between the city centre and its surrounding districts. The city centre has been a considerable economic success. Many companies have moved in, particularly companies active in new urban growth sectors such as ICT, financial services, etc. The city centre's high-level amenities attract many visitors who both make the centre lively and spend a great deal of money. Key magnets are the city's new large shopping centre, the concert venue and its many cultural facilities. Furthermore, the city centre has gained importance as a high-quality place to live: many old industrial buildings have been turned into up-market apartments, for which there is high demand. By 2001 the centre's population had grown to 6,100. One-third of the private dwellings were built over the three years before. The building explosion continues. The population was expected to reach the level of 13,000 by the end of 2003 (Manchester City Council, 2001e). The newcomers are generally well-to-do people. Their presence contributes to the liveliness of the city and also brings in much purchasing power.

Outside the city centre, the picture looks a lot less rosy. It is indicative that, of the city's 33 wards, 27 are in the top 10 per cent of the most deprived wards in the UK. Around 58 per cent of children under 16 live in low-income families, with this rising to 70 per cent in some wards (DTLR, 2000). A 2001 Council Paper links these grim facts to the need for e-government: 'These levels of poverty and social exclusion create a huge challenge for the local authority in delivering real e-government, whereby the local people have access to information, services and opportunities to become involved' (Manchester City Council, 2001a, p. 2).

Manchester's Regeneration Strategy

Manchester City Council has responded to the major urban challenges in various ways, but the general theme is reconciling economic regeneration with social inclusion. The greatest effort has been made in the large deprived area of East Manchester. This former industrial area has lost most of its jobs in the last decades and has high levels of unemployment. The quality of housing is low, crime rates are high and the education level of its inhabitants generally does not meet market needs. Many people have left the area in the last decades: the population has dropped from 60,000 to 30,000.

The regeneration strategy can be characterized as area based, integrated and opportunity oriented. It seeks to tackle social and economic issues simultaneously. The strategy has a timespan of around 15 years. Key projects include the development of up to 12,500 new homes, of which many are for higher income groups. The aim is to double the population to a total of 60,000 (Manchester City Council, 2001d). A new stadium was built in the area that was a major venue for the Commonwealth Games in 2002; nearby, retail outlets are being developed, specializing in sports-related items. The northern part of the area will become a business park. This park, at a distance of two miles from the city centre, should become the major employment base in the city, with 10,000 jobs. The area is easily accessible by motorway as well as public transport. A new public transport interchange is envisaged on the line of the Manchester to Oldham/Rochdale Metrolink extension. Many initiatives are being taken to empower local people and to tackle all kinds of social problems in the area. For instance, substantial efforts are being put into improving primary and secondary education in the area. The regeneration framework is being implemented by a limited company named New East Manchester Ltd (NEM). This company is a partnership between the City Council, English Partnerships and the North West Development Agency. Information and communications technology (ICT) is being incorporated as an important tool to achieve economic and social goals in East Manchester. ICT is seen as a tool to improve local services, to empower local communities, to address the skills mismatch and to improve public service delivery and safety. As we will see later in this chapter, the area even functions as laboratory for several local access and content policies. In the next section, however, we will first analyse Manchester's e-strategies.

3 Manchester's E-Government Strategies

Introduction

The city of Manchester has long recognized the importance of ICTs in a number of policy fields. In 1991 Manchester was the first local authority in the UK to develop an online public access information network. In the same year it launched a network of Electronic Village Halls to support community access to computers. Manchester was one of the founders of Telecities, a European network organization in which major European cities exchange experiences of ICT use for all kinds of purposes in the urban context. In recent years, developments have accelerated and the scope and impact of the city's ICT visions and strategies have widened. In this section, we will discuss Manchester's e-strategy. First, the national context is sketched in. This is extremely relevant since, in the UK, national e-government strategies have a major local impact and in many respects set a framework for local policies. Second, we describe Manchester's e-vision as laid out in the official Council Statement on e-government.

National ICT Policies: Content, Access and Infrastructure

Content
The UK's Labour government has e-government high on the agenda. A national e-government strategy was drawn up in the late 1990s. This strategy stressed the importance of ICTs, as instruments for economic and social development and as tools to improve public service delivery. The ambition of the UK government was to have all government services online by 2005. The government envisioned achieving this through 'joined-up' government: government agencies need to

cooperate and jointly build e-services that are of use to citizens. Partnership is a key word in the UK strategy.

National government plays an important role in setting the framework for local e-government. In April 2002 a national strategy for local e-government was published. This consultation paper, jointly drawn up by the Ministry for Local Government and the Regions and the Local Government Association, 'sets a framework of standards, expectations, infrastructure and support within which local innovation and delivery can flourish' (DTLR, 2002, p. 3). The proposed strategy is based on another White Paper, 'Strong Local Leadership – Quality Public Services', in which a strategy is laid out to improve local public services delivery in partnership models. Central government wants to see local strategic partnerships (LSPs) across the country, to establish integrated approaches for integrated services delivery. It has drawn up a list of 700 online services.

Thus, national government sets the institutional and legal framework for local e-government. However, for local e-strategies and projects, national e-government policy is important in at least three more respects. First, national government gives a number of financial incentives to make local councils active in e-government. An example is the Invest to Save Budget, under which cities can submit proposals (always in partnership with other organizations) to receive funding for e-government projects. In these proposals, cities have to outline their plans and also make explicit estimates of the costs savings and quality improvements they expect to achieve. In 2001, Manchester won an e-government project worth €1.8m to develop a web application for the maintenance supply chain for social housing. In 2002, a project worth €2.5m was granted for the development of transactional web services across the city, starting in East Manchester, to be extended to other areas in a later stage.

Second, national government has forced cities to think strategically about e-government. In order to get any national funding, every council had to produce an 'Implementing Electronic Government' statement, in which it sets out what its ambitions are, and how they are to be achieved. The Manchester City Council produced this document by July 2001 (Manchester City Council, 2001a). There will be annual feedback to look how far the ambitions have been realized. According to some council officials, the drafting of the document has helped the city council to replace its *ad hoc* e-government policies with a more integrated and comprehensive approach. It also raised awareness of the importance of e-government in departments that used to be more sceptical and/or passive.

Third, and more indirectly, national government has introduced a system of performance indicators. On an annual basis, cities must demonstrate to their inhabitants and to national government how well they have performed in a number of respects. This stress on performance puts more pressure on performance-increasing e-government projects. One of the problems for Manchester in this respect is the lack of (financial) resources. National government puts a lot of pressure on local councils, but in the eyes of some Council officials, fails to recognize the poor financial position of many councils.

Access and infrastructure

The UK has set itself a bold ambition: to be the best place in the world for e-business and e-commerce and a world leader in the electronic delivery of public services through e-government. The government has highlighted four key challenges which must be addressed if these ambitions are to be fully realized:

• developing a world-leading broadband infrastructure with access for all, from all businesses, especially SMEs, to everyone in the wider community;
• ensuring that the education and the skill base is there to develop and sustain the future workforce;
• tackling the 'digital divide' and ensuring that the information society is open to all;
• creating a business-friendly environment for e-commerce and e-business to develop and reach a critical mass, especially in terms of the digital content industry.

The broadband target was that, by 2005, the UK would have the most competitive broadband infrastructure of the G8 nations. A 'UK broadband fund' was established by the DTI (Department of Trade and Industry) to use public money to speed up the rollout of broadband to people and companies so as to achieve a leading competitive infrastructure. On a competitive basis, a fund was distributed to all the UK regions to demonstrate the benefits of broadband connectivity and applications. In the northwest, a regional bid led by the NWDA (North West Development Agency) invited bids from the five sub-regions of Cheshire, Cumbria, Lancashire, Merseyside and Greater Manchester to submit strategic proposals for deploying UK broadband funding in their area. The consultants SQW (2002) expected that in due course, demonstrator activity under the broadband fund would attract more substantive funding provided by the NWDA, national government and the European Union.

Regional ICT strategies There is also a regional ICT strategy, produced by the NWDA (2002), which highlights the importance of ICT developments for regional competitiveness and warns that unless the public, private and voluntary sectors seize the opportunities that ICT offers and join together in a strategic response, competitor regions at home and abroad will overtake the UK. Similarly the Core Cities working group on transport and telecommunications identifies the potential for city economies to gain competitive advantage from investment in knowledge based industries and services. Continued economic growth and sustainability will be increasingly dependent on the ability of city economies to support and sustain innovation and the knowledge base and to find ways of translating these into high-value products and services.

The Urban Level: Manchester's E-Vision

The ICT vision of the City of Manchester fits into the national and regional context. Manchester City Council's vision of is to make Manchester a leading world class digital city, with one of the most competitive broadband infrastructures in Europe, attracting and sustaining investment in ICT and e-commerce across all sectors of the economy, generating new businesses, developing new learning cultures, promoting social inclusion and providing all residents with the skills and aspirations to play a full role in the information society. Manchester's economic base needs to be modernized and equipped to compete more effectively in global markets, which are increasingly characterized by ever-greater usage of ICT and e-commerce.

The successful 'digital city' needs a new strategic framework into which four key information society themes (e-business/e-commerce, e-learning, e-communities and e-government) can be properly integrated and then developed in ways which will contribute most effectively to the City Council's and the local service providers' principal objectives in this field. These should include:

- promoting local employment opportunities through the Manchester Employment Plan and PSA (Public Service Agreement) employment target;
- ensuring that civic institutions, the education base and key development sites have state-of-the-art connectivity;
- helping decision-makers to develop a more sophisticated understanding of the leading edge of ICT developments and investment required to create market advantage in the knowledge economy.

To achieve this, the Manchester Digital Development Agency (MDDA) has been established, to ensure that Manchester has a properly 'joined up' approach to all aspects of ICT development and information society issues. Building on more than 10 years of pioneering work in this field, the time is now right for a new and more coordinated approach which will deliver the aims and objects set for the MDDA, including:

- aiding Manchester's transition to a knowledge-based economy;
- helping to generate high skill, high value jobs for local people;
- bringing together the various ICT strategies across the city and integrating these into one city-wide Manchester e-strategy;
- supporting and adding value to the work of the LSP by providing a single point of reference for the LSP's work on ICT and information society initiatives.

Thus, Manchester's vision on ICT includes elements of content, infrastructure and access. The vision is not only on paper: it has been translated into strategies and action plans in each of these fields. They are summarized in Figure 7.1.

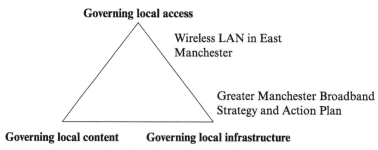

Electronic Village Halls
Access in schools
Cheap computers for poor households

Governing local access

Wireless LAN in East
Manchester

Greater Manchester Broadband
Strategy and Action Plan

Governing local content   **Governing local infrastructure**

Manchester's e-government strategy
Content for East Manchester: Eastserve.com

Figure 7.1   Content, access and infrastructure strategies in Manchester: An overview

In the field of access, ensuring access to computers and the Internet for all citizens remains a priority. Access policy is not new for Manchester. The city was even a frontrunner (with its Electronic Village Halls already established in the early 1990s). The emphasis has now changed to include access in the home as well as broadband access. In section 6, we will discuss access strategies in more detail. As we shall see, huge efforts are made to empower people in deprived areas with the help of ICT tools and skills. In the field of infrastructure, the focus lies on deprived areas as well: the Council has ambitious plans to construct a wireless wide area network to provide wireless Internet access for inhabitants of East Manchester. Also, the city is involved in the Greater Manchester Broadband Strategy and Action plan which was drawn up in the summer of 2002. We will discuss this strategy in section 7.

In the field of content, the e-vision encompasses a number of ambitions. First, the city plans to use ICT to enhance the internal organization. To that end, it will implement enterprise resource planning (ERP), in strategic partnership with ICL/Fujitsu and Deloitte&Touche, to consolidate financial information systems, and consolidate data and the use of data warehouses (more on this in section 5). Furthermore, it embraces ICT to improve external relations with both customers and suppliers. It will use advanced call centre technology and web applications to facilitate transactions and community portals will be developed to provide an effective interface between customer needs and Council information systems. E-procurement will make procurement more efficient. In its strategy document, the Council states that partnership (with public and private partners) is key to coordinate the provision of content and to deliver seamless services.

4   Governing Content

Introduction

In this section, we will describe and analyse in more detail how the City of Manchester seeks to put electronic services online and the way it deals with the internal organizational consequences. We will also discuss Manchester's presence on the Internet.

Public Content: The Provision of Local E-Government Services in Manchester

For the City of Manchester, improving service delivery for the urban community is an important policy target. ICT is a key tool in this respect.

The City Council consists of a number of departments, each with its own IT group. The central IT unit is part of the Corporate Services Department (formerly the City Treasurer's Department). Its role is to design/maintain the overall IT architecture and to control the central systems and databases. The IT groups in other departments can purchase their own systems if they like, but they always need approval of the central IT unit. Furthermore, the central IT department acts as contracting agency. The desktop computers and their maintenance for the entire City Council are contracted out to Compaq. British Telecom (together with Nortel and SCC) currently rolls out a broadband network linking several council offices. The IT department negotiates and manages these contracts.

The central IT department plays a key role in the improvement of customer relation management (CRM). For instance, it is heavily involved in the set up of a pan-municipal call centre that will allow citizens to have a single contact point for all their questions. Regarding the municipal website (www.manchester.gov.uk), the role of the IT department is primarily to provide technical support. Following a tender process, the city has appointed 'Your Communications', (formerly Norweb Telecoms) as the Internet service provider. Each individual department publishes its own content on its part of the municipal site. Currently, the overall responsibility for the website is at the Chief Executive office. In the near future, the IT department will take over this role.

The city of Manchester's aim was to have all its service delivery online by 2005, in line with government targets for all public bodies in the UK. The Council has worked hard to achieve this. Online service delivery is part of a broader strategy to improve the service performance of the Council. The city has decided not to do everything at once – it lacks the resources to do so anyway – but to focus on services where the marginal returns in terms of benefits for the citizens are the greatest. Until the time of writing it had put priority on e-services in the field of housing – the Council is a substantial landlord, owning 60,000 flats) and on information on welfare and other benefits. Other e-services will be dealt with later.

The Housing Department is the most advanced. It has assembled an online directory with all the 60,000 council flats. On the website, a 'home finder programme' enables users to search and subscribe to houses, after typing in a number of search criteria. The site yields pictures of the homes and characteristics, as well as information on the neighbourhood. As a second feature of the site, tenants can report faults and book a date for repairs. The system links up with systems and agendas of suppliers (plumbers, carpenters, etc.). 93 per cent of all possible types of repairs are included on the site. This project was financially supported by national government (€1.8m from the Invest to Save scheme). Finally, the site offers a tool to calculate the rent and the eventual benefits for tenants. This used to be a complex task, given the myriad of rules and laws: now, the computer does the work.

It took the housing department one and a half years to reach where they are now: two employees have worked on it on a full-time basis. In the setting up, the most difficult thing was to analyse the manual processes related to the services and to arrange the interaction with the old back-end IT systems. According to officials from the Housing Department, the service is now widely used and saves a lot of paperwork for the Department. In the near future, the 'home finder' may be extended to include non council-owned properties as well. Other departments in the municipality lag far behind the Housing Department which, because of its experience, now acts as advisor to other municipal departments for introducing web applications.

At the time of writing the Council is in the process of setting priorities for the further development of e-services. The available resources do not allow doing everything at once, as huge investment costs are needed to implement the services.

Other departments of Manchester City Council have also developed web content. Each of the Council departments is responsible for the publishing of its own content. This results in large qualitative differences between the departmental web pages. Overall responsibility for the website is formally with the 'central team' at the Chief Executive's Office. In practice, there is no strong central control on content, but rather on the use of the correct standards and logos. The lack of interdepartmental coordination leads to inefficiencies. In the near future, responsibility for the content of the council website will move to the central IT unit.

Towards Integrated Services: East Manchester as a Test Laboratory

In the deprived area of East Manchester, a large regeneration programme is being implemented (see section 2) in which ICT is being incorporated as a tool to achieve things. The programme brings substantial financial means and other resources into the area. From this perspective, East Manchester functions as 'test laboratory' for e-government solutions. In 2000, an ambitious website – Eastserve. com – was launched in the area. This is an information portal enabling people to access services from seven agencies, among which are the City Council, health and education providers, the police and voluntary sector bodies. The website includes a job database, an application to calculate entitlement to and amount of benefits; local community information; discussions forums; and local news. It incorporates online housing application forms and a facility to request repairs. Citizens can also use the website to report crimes to the police. One of the problems of the services proved to be the difficulty in keeping all the information up-to-date. For instance, the police used the site to show pictures of missing persons, but in some cases failed to withdraw them when a person was found.

At the time of writing, the portal is being upgraded to include more transactional features. The underlying vision is to deliver online a wide variety of services to local residents in a manner that is appropriate to their situation and need and which joins services up to suit individual customers and citizens (Manchester City Council, 2001c). In other words, this is a service focused on local demand, not on the supplying agencies. Among other things, the portal will include the ability to make payments (ranging from parking fines to commercial rents); offer online advice; and reformulate information for specific target groups. Transactional services in the fields of education, police, health and housing will be improved as well. The management of the content will change also. A content manager will probably be appointed. This person will become responsible for the updating of content from all the different sources. Also, efforts are being made to 'regionalize' some databases. For instance, currently the website has a link to a UK-wide job openings database ('worktrain'). In the new system, only the jobs in the Manchester area will be shown.

The City Council expects substantial benefits from this initiative, not only in terms of improved service delivery for the citizens of the area but also in terms of cost savings. A total cost saving of €4.2m over a five-year period is expected. This sum is made up of a reduced number of calls to the housing call centre, fewer manual processing of payments, a reduction of back-office and support costs at the

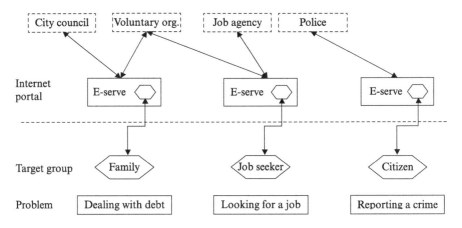

Figure 7.2   Eastserve.com

Council and the possibility of pooling resources among the seven organizations that are involved in the Eastserve project (Manchester City Council, 2001c). In April 2002, a total of 27,741 people visited the website (TNS/C&L stat server).

Manchester City Council, as lead partner, is working in partnership with a private company which is setting up the technical platforms and with New East Manchester Ltd, the public company established to coordinate the regeneration of East Manchester acting as the key management agency.[1] Some large multinationals contribute in kind to the project. Cisco systems provided some equipment and Oracle offered free Dbase systems.

As already mentioned, the project is intended to be citizen-oriented. To that end, community and voluntary organizations have been deeply involved in the set-up and implementation of the project. From an organizational perspective, the key challenge will be to build a content architecture which can deliver content from several agencies in a comprehensive, citizen-oriented way.

The initial implementation in East Manchester will be used to demonstrate the ability of ICT to support improvements in the quality of service to residents and to reduce the cost of transactions. After implementation in East Manchester, services will be rolled out to two other regeneration areas (North Manchester and Wythenshawe). They will all bring in their own local colour. Ironically, the least favoured areas are the first to be endowed with integrated, demand-oriented online public services.

It is important to note that Manchester closely involves the users and experts in its e-government initiatives, not only in East Manchester but also elsewhere. The city involves representatives of the community and voluntary organizations, alongside the private sector, other local public agencies and the academic sector, in 'sounding boards' to test the e-government approaches (Manchester City Council, 2001a, p. 13).

Rearranging the Back Office: The ICL/Fujitsu/D&T Partnership

The City Council is aware that implementing e-government requires both new organizational concepts and integration of IT systems. This is a very complex operation which needed specific know-how that was not available within the organization. Therefore, in 2000 the city advertised for a partnership with private business to improve local services. Consortia could bid for the job. Several bids came in and, ultimately, the combination of ICL/Fujitsu (a computer company) and Deloitte&Touche Consulting won. They proposed to combine a business process reengineering implementation alongside a rearrangement of IT systems. After the consortium had won the bid, a 'principle agreement' was signed, after which negotiations over the details started. This proved to be a difficult trajectory,

---

[1] 25 per cent of the costs (€3.16m) are borne by the City Council: the other 75 per cent is funded through the national ISB programme.

with long and difficult rounds of negotiations. Both parties found it hard to give substance to a real 'strategic partnership' between city and the businesses. Up to the time of writing (May 2002), no formal final agreement had been signed. It has become clear that the cooperation will take the form of a risk/reward strategy: the consortium will be rewarded when results are good but also carries part of the risk of failure. The time span of the cooperation will be seven to 10 years. For the Council, engaging in this kind of partnership is new. Some pilot projects have already been undertaken. In the areas of benefits, home care and public works, some re-engineering trajectory pilots have been undertaken. Results or outcomes, however, are not publicly available.

In the process, the relationship between the consortium, the central IT unit and the IT groups in the other Council departments has been troublesome at times. Most curiously, the former head of the IT department was not involved in the Business Process Reengineering (BPR) project. However, this has changed with the newly-appointed head. Also, in some instances the departments work together with the consortium without consulting or involving the central IT unit.

It is clear that, whatever the shape of the partnership will eventually be, consequences for the Council organization are considerable. In the end, the quality of the services will depend on the quality of the civil officers and the organization. For one thing, customer care will become a more central concept. Council officers need to learn to think more in terms of the customer. Furthermore, cross-departmental cooperation will need significant improvement, as integrated e-services are typically composed of content from different departments. This requires a considerable cultural change within the organization.

Promoting Local Content Production

The City Council not only deploys ICT in its own organization: it also actively supports the use of ICT by local communities. This is achieved through the Manchester Community Information Network (MCIN) organization. This project is funded through the Regeneration Partnership (in which the City Council is key partner) and European Structural Funds.[2] One of the organization's objectives is to increase awareness of the Internet as a means for grassroots organizations to achieve their goals. It helps and supports individuals and community groups to go online by providing them with training and a basic content editing software package. Furthermore, MCIN brings all the initiatives together on one community portal site, MyManchester.net. The portal is visited frequently and visitor numbers are increasing, from 2 million in 2000 to 2.5 million visitors in 2001. At the time of writing, there are the extensive web activities of local communities and volunteer organizations. In total, 300 of this type of website have been set up in the last few years. They service a rich variety of purposes and target groups: some are area-

---

[2]   The total annual budget of MCIN is €395,500.

based, others are interest-based communities. For instance, there is a platform on mental health and one for black communities. Local grassroots organizations, or enthusiastic individuals have set up the websites.

Organizing Local Content

There are many different sources of local digital content: public bodies (like the City Council), semi-public institutions, community groups and private companies. For users of digital information/services, it is convenient if content from various sources is organized in a demand- or user-oriented way.

In Manchester, we can discern a number of local website categories for different user groups. The first category concerns the integrated portals for local communities that we just described. These websites are primarily local, aimed at citizens in a specific borough. They bring together services of different public agencies and grassroots organizations. These are the most advanced websites in terms of interactivity. In the case of the East Manchester organization, the local public regeneration company is responsible for content management. It steers the suppliers of information, etc. This approach can be characterized as demand-oriented and integrative.

A second user group, investors and companies, is serviced online through manchestercalling.com. This is a portal website, containing everything you need to know about Manchester as a business and investment location. The site is owned and managed by MIDAS (Manchester Investment and Development Agency Service), part of Manchester Enterprises Ltd. The site offers a variety of information and links about workforce, transport and communication, education, property and sites, but also about lifestyle, the arts, leisure, tourism and living in Manchester (house prices, taxes, etc.).

For a third user group, tourists, several portal-like websites are available. There is the 'destination Manchester' website, produced by Marketing Manchester in partnership with the Greater Manchester local authorities (Bolton, Bury, Manchester, Oldham, Rochdale, Salford, Stockport, Tameside, Trafford and Wigan). It claims to be a fully-fledged tourism portal for Greater Manchester, offering the very latest information about events and attractions. You can also browse online for top quality accommodation and travel advice. However, the content of the site needs improvement: the number of hotels and tourists attractions in the website's database is too low to match its pretensions. In the near future, the organization will become more proactive: it will produce a regular Destination Manchester newsletter totally tailored to users' needs and featuring offers and events.

There are also commercial websites offering tourist information on the city. The most extensive one is 'Virtual Manchester', to be found at www.manchester.com. It offers 12,500 pages with comprehensive information for tourists – among other things, interactive maps, public transport timetables and tourist guides – including interactive booking facilities. It also contains many links to commercial businesses in the city. The website is heavily sponsored. Several search engines give the site high grades for its interactive possibilities and comprehensiveness of information. According to the advertising agency that sells advertising space on the website, the site attracts an impressive 2 million hits per week (104 million visitors on annual basis).

(The Lack of) Broadband Content

As in most other European cities, Manchester's development of broadband content – services that need high speed-transmission for effective delivery – is in its infancy. This is concluded by the Greater Manchester Broadband steering group which commissioned a study to assess the local broadband situation in the area in terms of access, infrastructure and content and to formulate ambitions. A report by consultants SQW (2002) concluded that one of the reasons why the take-up of broadband is relatively low is the lack of perceived added value: consumers simply do not see why they should pay more for more bandwidth. Several actions are suggested to stimulate the development of broadband content. First, the development of services, structures and processes which are most effectively accessed through use of broadband channels. Public sector organizations have a key role to play in 'leading by example'. However, in Manchester, until now there are no examples of public broadband services.

Second, the establishment of initiatives and partnerships at local level was advised, to deliver inter-agency working and provide a joined-up public sector offering, delivered by broadband, to citizens. Third, the report encouraged engagement with Greater Manchester's large digital industries sector and existing content providers, to gain buy-in to the strategy, exploit local expertise and capture regional and wider funding support for developing relevant and beneficial broadband services for Greater Manchester's broadband users. This seems to be a promising road, as the city has substantial 'creative energy' embedded in its cultural and ICT industries (see van den Berg, Braun and Van Winden, 1999).

5   Governing Access

Access Policy has a History in Manchester

The city of Manchester has a longstanding tradition in policies to promote the use of ICTs, particularly by less favoured groups and in public places. In the early 1990s 'Electronic Village Halls' were already established. These were public places where people could use computers and access council information. In the course of the 1990s, access policies became stronger and broader. The council realized that access to ICT – both in terms of skills, computer possession and Internet access – was increasingly crucial for full participation in the information society, both in social and economic terms. By the end of the 1990s, the city had a number of public or semi-public places where citizens could access the Internet and many initiatives had sprung up to teach people all kinds of computer skills. All public libraries are equipped with computers: the Electronic Village Halls offer courses for specific groups (women, ethnic minorities) and several ICT learning centres have been established. This latter initiative was partly supported by national government. There are also several grassroots initiatives aimed at improving access: for instance, in East Manchester a church minister has set up an ICT-education scheme for children who have left school, as well as a centre for ICT education. ICL (a computer manufacturer) has provided the equipment. By 2005, the city was aiming to have up to 100 public kiosks for accessing services online (Manchester City Council, 2001a). The area of East Manchester is best endowed with access facilities and will become even more so in the near future. ICT is regarded as an important element in the regeneration of this very deprived area. We have seen that this area is a pilot area for 'rich local content' developments: the same holds in the field of access. The area counts several public places for Internet access and computer use as well as training centres.

Shifting Focus: Individual Access and Broadband Access

In the access policies, a shift in focus can be observed towards individual access (computers and Internet in the home) and broadband access.

A relatively new initiative includes policies to promote individual ownership of devices and home access. In a particular area in East Manchester, citizens can get a brand new Internet-ready computer with monitor and colour printer for €317, a recycled computer[3] for €48 or even less for people on benefits, a free Internet-only computer for €80 or a free set-top box (enabling Internet access via TV). On average, the hardware is subsidized for an amount between €317 and €476 each.[4] The project has a long timescale and started only recently. Nevertheless, first indications show that few people in the area are interested in the devices. In the target area of 4,500 households, only 1,250 devices have been sold. There may be several reasons for this low take-up. First, some people in the area already may have a computer (although this group can be expected to be small). Second, some groups of people will never want one, for instance the elderly, because they do not see the value added, or simply because they fear new technologies. It can also be the case that even the low prices for the equipment are too high for some groups in the area. Finally, our interviews suggest that some people may feel distrust towards the City Council's good intensions with the project. They may erroneously think, for instance, that the council uses the equipment to check their behaviour. Further research should reveal the reasons of the low sales levels.

---

[3]   These computers are assembled by unemployed people in East Manchester through the ITEM project.

[4]   Other UK cities have similar initiatives. In Blackburn, devices are given away.

Second, access to broadband is considered an important issue. Recently, a Greater Manchester Broadband Steering Group commissioned a Greater Manchester Broadband Strategy and Action Plan (SQW, 2002), in which promotion of broadband access – both in the home and in public places – was strongly recommended. The aspiration level was formulated as follows: 'Greater Manchester aspires to be a sub-region in which all citizens, communities and businesses have at least one place where they can actually use broadband[5] services and come into contact with the benefits'. Partly because of the high number of public places where the Internet can be accessed for nothing, Manchester has only very few Internet cafés.

6 Governing Infrastructure

This section deals with the local electronic infrastructure situation in Manchester, as well as with ICT infrastructure policies in the city and the region.

The ICT Infrastructure Situation in Manchester and the City Region

Not all neighbourhoods in the city of Manchester are equally endowed with electronic infrastructure. The best coverage is offered by the traditional telephone landline. This means that, in principle, every citizen in the city can have dial-up access using this network. However, in practice, in several problem neighbourhoods, some people do not have a landline connection. In East Manchester, 25 per cent of homes do not have landlines (SQW 2002, p. 9). Large groups in these areas are relatively poor and have difficulties making ends meet. Some people failed to pay their bills and were cut off by the telephone operator. Many people have switched to cheap 'pay as you go' models of mobile telephony.

Broadband is less widespread. Although technical ADSL coverage (by British Telecom) in Greater Manchester stands at 97 per cent (SQW 2002, p. 9), broadband penetration to the home is very low in Greater Manchester: by April 2002, only 1.16 per cent of Greater Manchester homes were ADSL subscribers. There is very limited competition in the broadband market. Higher bandwidth connectivity is readily available from a number of carriers such as Your Communications or Kingston Communications, but British Telecom dominates the market to a large extent. The cable network covers only 30 per cent of the Greater Manchester Area. In some parts of the metropolitan area, the cable network is extensive. All the richer or middle-class boroughs are 100 per cent cabled; the same holds for the problem neighbourhood of Wythenshawe. NTL is the dominant cable company. Other areas, such as East Manchester, have no cable installed.

---

[5] Broadband is defined as 2Mbps upstream and downstream at minimum.

Broadband Policies

In the UK, the construction and operation of electronic infrastructure provision is generally regarded to be a task of the private sector. Nevertheless, public sector bodies promote infrastructure developments in several ways. The national government has the ambition of making the UK the most competitive broadband market by 2005 (SQW, 2002). In 2001 North West Region produced a document entitled 'englandsnorthwest connected', in which it announced actions to achieve a world-class broadband infrastructure. The report makes mention of significant geographical pockets of weakness, not only in rural parts of the region but also within metropolitan areas (NDA 2001, p. 19).

Broadband to schools and community centres

The National Grid for Learning Programme is stimulating the connection of schools, with all secondary schools to be connected by September 2002 and primary schools to follow. Post-school, Greater Manchester has 3 LearnDirect Hubs, connected to around 100 learning venues. Net North West embraces the powerful and well distributed dark fibre network established in Manchester – G-MING (the Greater Manchester Inter-Networking Group), provides the connection to the national academic backbone SuperJanet4, manages high bandwidth connections to the universities, higher and further education providers and many schools in Greater Manchester, including many of the 25 Learndirect Centres, has high bandwidth connections into many of the hospitals and research centres and connects many libraries and other public resources. Collectively these learning and other places have a presence in all communities and their network providers have both spare capacity and a commitment to support the communities they serve.

Wireless broadband

In a very ambitious scheme, the City Council is now providing the area of East Manchester with a wireless broadband wide area network (WAN). This will enable people in the area (4,500 households) free access to the local intranet. This intranet will contain several services, among which is the Eastserve.com portal discussed in section 5. People who want to access the Internet need to pay the normal monthly fee to an Internet service provider of around €24. Not only houses will be connected, but also community centres, schools and several public Internet access points.

In the roll out of the network, these will be the first to be connected. A private company will effect the roll out. The maximum speed will be 10Mbps. This is a high speed at almost no cost. A 'normal' ADSL user in the UK gets a speed of 512Kbs for a monthly fee of around €40.

The idea behind the project is to give excluded groups access to broadband Internet. Broadband enables citizens to access and use rich content like, for instance, moving images. Broadband enables more personalized forms of online service provision, for instance using videoconferencing technology. Furthermore, it enables more advanced types of online learning.

Typically, a wireless broadband WAN is expensive to build, but once in place, operational costs are very low. In this case, capital investment amounts to €3.2m. Figure 7.3 shows the envisioned configuration of the network.

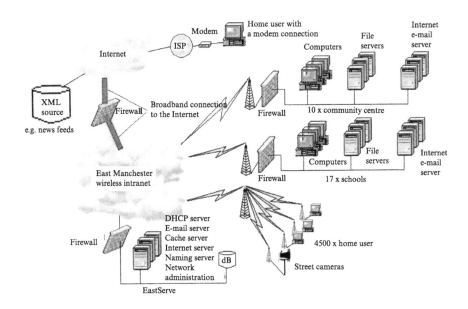

Figure 7.3   Configuration of the network

7   Conclusions

Manchester's ambitions as laid down in its e-strategies are high and many efforts and resources are being invested to turn the ambitions into reality. The vision and strategies regarding ICT are very well aligned with the general urban strategies to combine urban economic revitalization with social inclusion policies and to improve levels of service provision. National e-government policies strongly encourage a speed-up of ICT policy implementation in Manchester by providing institutional frameworks and making funds available.

In our frame of analysis we make the distinction between access, content and infrastructure strategies. Access strategies concern actions to include everyone in the information society. Content strategies include the provision of useful electronic services that improve or complement traditional services. Infrastructure strategies are directed to improving the quality of the local electronic infrastructure. For each of these domains, we have assessed the ability of 'local organizing capacity' to get things done. Organizing capacity contains the following elements: vision, strategy, public-private networks and political and societal support. It is embedded in a spatial economic and cultural context that is domain and urban specific.

What can we say of Manchester's e-vision and strategies on the basis of this framework? In the field of ICT access strategies, Manchester has a strong tradition on which it continues to build. The city was one of the first in Europe to open public Internet access points in the 1990s,and is now a frontrunner in promoting individual computer ownership and (broadband) access in deprived areas. The importance of access to new technologies for all is recognized throughout the City Council (political support seems to be strong), and resources are invested accordingly, particularly the area of East Manchester but also in other places. The efforts reflect the idea that ICT can make a real contribution to social inclusion of less advantaged groups. For a city like Manchester, with severe social problems, much is at stake to make these policies a success. In a way, the city is a 'test lab' for the effectiveness and efficiency of 'social inclusion-oriented' ICT strategies. Some experiments may be considered a little extreme (one example is the newly-developed wireless broadband network in East Manchester), but they may reveal very relevant lessons that can be applied in other boroughs of Manchester or in other cities.

The 'access policies' in Manchester are designed and implemented by a combination of the City Council, the voluntary sector and all kinds of grassroots organizations. This greatly increases the chances of success and probably contributes to the flexibility of policies. We feel that private sector participation in ICT access policies could be a point of improvement. Recent Euricur research (van den Berg, Braun and Otgaar, 2002) revealed a tendency for private enterprise to be prepared to invest more in social projects, from a 'corporate social responsibility' point of view. Structured efforts by the city could help to capitalize on this new opportunity. As a start, the city could organize a roundtable with the large (ICT) companies in the region to discuss the contribution of companies to regeneration projects.

Manchester is not only active in access policies but also strongly promotes the development of electronic content. In this domain, a distinction can be made between *suppliers* of content (public organizations, community organizations and companies), and *demanders* of content (various categories of citizens, firms, investors, tourists, etc.). The more the production of content is demand-driven, the higher the value added is for the users.

One of Manchester's key ambitions (like that of so many other cities) is to use ICT to improve the quality and cost-effectiveness of service delivery. Leadership in the field of implementing e-government has been weak and dispersed, but positive changes seem underway. In the implementation of the vision, leadership should be a combined effort of leaders in IT (in the City Treasurer's Department), the Chief Executive and the Economic Initiatives Group.

Internally, the city works hard to 'get the basics in order'. In cooperation with ICL/Fujitsu and Deloitte&Touche Consulting, a business-reengineering program is being implemented, to integrate IT systems and databases and align work processes and information flows. The city has not simply subcontracted the job, but engaged in a strategic partnership in which the risks and rewards are distributed between the city and the consortium. This reflects the fact that ICT implementations have become strategic: ICT touches the core business of the city and substantially affects the quality of its service provision. At the same time, cities do not have all the necessary high-level expertise in-house and need to involve professional companies. In this complex environment, simply 'outsourcing' is no longer an option. The Manchester approach will certainly yield key lessons for other European cities that are struggling with similar problems. For future research, it would be interesting to undertake an in-depth comparative analysis of strategic partnerships in e-government strategies.

Concerning web content, at the time of writing individual departments of the City Council develop web services. An example is the Housing Department's home-finder application. At the same time, the city is developing and promoting demand-oriented, integrated online services for target groups. These portals are composed of content from several agencies. For citizens, the City Council has high ambitions to arrive at demand-oriented web portals that include information and service from police, health, housing, education and voluntary and community organizations. The city has taken an unusual but very interesting approach by starting at the local level. Each area in the city will have its own portal, with similar services but with a distinct local colour according to the characteristics and needs of that particular area. East Manchester is the pilot area for the development of these innovative applications. The question will be how to transplant successful applications to other areas. Will there be a central 'content management organization', that offers integrated services in a demand-oriented way for the entire city? Or will there be a 'content manager' in every borough, responding to the individual needs and characteristics of that borough? This latter option seems more costly and the value added is questionable. The sustainability of the projects is a big issue. The projects in East-Manchester have big initial funding (ISB 3 and 4, European Funds), but what if funds end? Perhaps here also large corporations could be involved from a social responsibility angle.

For tourists, the Council supports a web portal as well: DestinationManchester. com. To create and maintain this portal, the City Council cooperates with other councils in the metropolitan area and with Marketing Manchester. In our view, it is a good idea to use the Internet as a tool to inform and attract tourists, in regional cooperation. However, performance could be improved substantially if the databases (hotels, things to see, transport and traffic information) were more substantial. The credibility of the portal is undermined as visitors to the site only see part of the total tourist offer of the region. We recommend a critical review of the website content, completion of the database and ensuring appropriate maintenance. For business and investors there is another portal: 'Manchestercalling' which in our view is very good, useful and user-friendly.

Regarding electronic infrastructure, the City Council is far from passive. In this chapter we have focused on one of the most remarkable initiatives in Manchester: the introduction of the wireless broadband wide area network (WAN) in East Manchester. This network brings broadband Internet access to disadvantaged groups in the area, even to those who do not have a telephone landline connection. This project can be evaluated in the light of the heated discussion about the use and necessity of government intervention in the provision of Internet access. Proponents of government intervention argue that there are still large groups of 'unconnected' people without Internet access at all. Furthermore, a broadband divide emerges between rich and poor people (as broadband is expensive) and between areas with and areas without broadband infrastructure (because of cherry-picking behaviour of telecom companies). People without (broadband) access are assumed to be worse off because they cannot benefit from the kinds of new opportunities that the Internet offers. They (and their children) also run the risk of lagging behind in terms of the computer skills that are needed in the information economy. Therefore, the government should invest in getting 'everyone online' or even 'everyone on broadband'. Proponents of policy intervention additionally argue that ICT adoption can be of strategic importance for urban development in many respects. The returns of policy intervention to speed up adoption may, under some conditions, be high. What we can learn from the adoption literature is that efficient policies should preferably not be generic but targeted at non-adopting groups (see van den Berg and van Winden, 2002; van Winden, 2001).

Critics argue that the dissemination of (broadband) Internet access will be rapid: like any other new product, early adopters get it first but are soon followed by larger groups. After some time, laggard groups (generally those with low incomes and low education) join as well. In this 'natural' S-shaped diffusion process there is no need for government intervention. Furthermore, government intervention in technology adoption bears risks: it may distort markets, as it tends to favour a particular technology (although this can be avoided using demand subsidies); also, in the future it may discourage the private sector from investing (for a full account of all the arguments, see Leighton, 2001).

In this case, how do the investment costs of the project relate to the benefits that can be expected? Narrowband dialup access offers sufficient bandwidth for the e-government services currently available, but not everyone has it. This project intends to bring the services to everyone, including those without landlines. It can be seen as an experiment of which the outcomes are uncertain by definition: this has never been tried anywhere else. Furthermore, there may be alternative benefits of the network: more highly-skilled people – for instance, young professionals who cannot afford to live in Manchester's city centre but like to live near it – may be more willing to move to East Manchester when a superior and cheap wireless infrastructure is in place. In other words, the project may contribute to the regeneration and diversification of area. We recommend very close monitoring of the use of the network, particularly of people without landlines.

Broadband policy is a big issue in Manchester. The Greater Manchester Broadband Strategy announced all kind of measures to promote the widespread adoption of broadband. At the time of writing, however, broadband take-up is low because of high prices and a lack of useful broadband content. In this latter domain, the Manchester City Council could do something. It could consider exploring the possibility of providing e-government services, which require high bandwidth, or encouraging other (semi-)public actors to do so. This type of 'leadership by example' would increase the credibility of the city's ambitions and contribute to the solution of the 'chicken and egg' problem.

More generally, we found that the activities in East Manchester are an excellent example of combination of access, content and infrastructure policy. Integrated online services are offered, alongside computer/Internet access and skills for many people. Testing innovative e-services in deprived neighbourhoods with very low IT take-up levels can be justified, as inhabitants of deprived neighbourhoods are relatively heavy users of council services (police, benefits, housing). It is a form of social policy to improve service levels first where the needs are greatest. It is too early to report concrete results, as the interactive features of the portal have are only just beginning to be developed. Here also, close monitoring of the use of the services is recommended.

References

DTLR (2000), *Index of Multiple Deprivation 2000*, Department of Transport, Local Government and the Regions, London.

DTLR (2002), *E-gov@local: Towards a National Strategy for Local E-government*, Department of Transport, Local Government and the Regions, London.

Leighton, W.A. (2001), *Broadband Deployment and the Digital Divide: A Primer*, Policy Analysis No. 410, August 7.

Manchester City Council (2001a), *Implementing Electronic Government*.

Manchester City Council (2001b), 'Invest to Save bid', round 3.

Manchester City Council (2001c), 'Invest to Save bid', round 4.

Manchester City Council (2001d), *Manchester Update*, 6 (1).

Manchester City Council (2001e), *Manchester Update*, 6 (2).

NWDA (2002), 'englandsnorthwest connected – An IT Strategy for the Region' (Ref 0083), http://www.nwda.co.uk/inside/bookshop.asp.

SQW (2002), 'SQW's ICT Expertise Helping to Bridge the "Digital Divide"', September, http://www.sqw.co.uk/latestnews/news2002.php?area=l.

van den Berg, L. and van Winden, W. (2002), 'Should Cities Help their Citizens to Adopt ICTs? On ICT Adoption Policies in European Cities', *Environment and Planning C*, 20 (2), pp. 263–79.

van den Berg, L., Braun, E. and Otgaar, A.H.J. (2002), *City and Enterprise: From Common Interest to Joint Initiatives*, Euricur, Rotterdam.

van den Berg, L., Braun, E. and van Winden, W. (1999), 'Growth Clusters in European Metropolitan Areas', *Eurocities Magazine*, 11, pp. 14–16.

van Winden, W. (2001), 'The End of Social Exclusion? On Information Technology Policy as a Key to Social Inclusion in Large European Cities', *Regional Studies*, 35 (9), pp. 861–77.

Interview Partners

Mr Carter, Economic Initiatives Group, Manchester City Council.

Mr Copitch, Director, Manchester Communities Information Network.

Mrs Martin, Lead Officer on E-government Strategy, Manchester City Council.

Mr Mather, Director Employment Training and Business Development, Leader on E-government and Community Based Regeneration, East Manchester Urban Regeneration Company.

Mrs Smith, Business Analyst and Lead Officer on E-government Implementation, City Treasurer's Department, Manchester City Council.

Mr Sumner, Director, Clicks and Links.

Chapter 8

The Case of Tampere

1 Introduction

In this chapter, we will describe and analyse Tampere's e-governance strategy. This strategy has been labelled 'e-Tampere'. In comparison with the other case studies in our analysis, e-Tampere is a special case. It has a mixed technological and societal angle: information society technologies. From the very start, however, business development and ICT cluster stimulation have been the main targets of the strategy. Some parts of the e-Tampere strategy pay due attention to the issues of content, access and infrastructure which we have identified in our framework of analysis. But following the framework in the description of e-Tampere would result in an analysis of the strategy that would be too narrow. Instead, we will follow the main structure of the e-Tampere strategy. We will start in section 2 with a general context description of the challenges that the city of Tampere faces. This puts the city's e-governance efforts into perspective. In section 3, we introduce the e-Tampere programme. In sections 4, 5 and 6 we analyse the main action lines of the e-Tampere programme:

* strengthening the knowledge base of research and education;
* creating and accelerating the growth of business activities;
* making public services available online.

Section 7 evaluates the e-Tampere programme.

2 Tampere: Profile of the City

The city of Tampere was founded in 1779. With almost 200,000 inhabitants in the city and 300,000 inhabitants in the Pirkanmaa urban region, it is the second largest regional centre in Finland. The regional industry is based on mechanical engineering and automation, electro-technical industry, information and communication technology, biomedical technology, pulp and paper industry and graphical industry.

The city of Nokia is situated in the Tampere urban region. It was the birthplace of Nokia enterprises, which over the years developed from a forestry-related industry through different stages and sectors of industrial production into a world-leading mobile communications firm. This development illustrates the overall development of Finland and the Tampere urban region. The United Nations

Human Development Report 2001 classifies Finland as the most technologically advanced country in the world. Finland leads the world in Internet services, export of high technology, numbers of students enrolled for technology studies, etc. (UNDR, 2001).

Tampere's Municipal Policy for the Twenty-first Century

In 1997 the City Council of Tampere endorsed the Tampere City Strategy programme (City of Tampere, 1997). This programme defines the guidelines for developing Tampere in order to maintain the prosperity and well being of the city, its inhabitants and its companies. The city's prime responsibility – according to this programme – is to provide residents with the conditions for a good quality of life. A first element is to maintain the city's position as a leading industrial city in Finland.

At that time the city of Tampere had 186,000 inhabitants. It was expecting an increasing demand for social and health services. Unemployment was high. The share of elderly people in the population of Tampere was increasing at a slightly higher rate than the EU average. These facts constituted some of the major challenges for the local governments in the Tampere region. The information society technologies and other knowledge intensive activities were clearly identified as ways to meet the challenge. Growth of the knowledge-intensive sectors would attract highly-educated young, working-age residents to the city. The already-existing presence of two universities and two polytechnics – with their respective spin-offs in terms of business development – would certainly add to the potential.

The Tampere City Strategy programme defines the guidelines for developing Tampere. The programme states that a good quality of life and the financial well-being of the citizens are to be based on thriving industries and well-balanced public finances. The production of *information technology applications* for industry, education, administration and recreation is identified as one of the growth sectors for the regional industry. The promise to residents is that they will have *equal access to the information society*. They will benefit from improved *information network infrastructures* and new *municipal services* available on these networks. Thus, already at this early stage the municipal strategy addressed the issues of content (including application services), access and infrastructure that we use as a reference model for urban e-governance strategies.

However, within the Finnish national urban system, there is strong competition from Helsinki. Attracted by the opportunities in the capital region, many young people – including some youngsters from Tampere – prefer to move to Helsinki (Kasvio, 2001). The e-Tampere programme is a clear expression of the local government's intention to cope with this kind of observation.

The overall strategy for Tampere was updated in October 2000, when the City Council endorsed a plan that outlines the development of the city in 2001–2012 (City of Tampere, 2001). This plan builds on the strategic approach of the 1990s while taking account of the downward slope of the 'dot.com economy' at the turn of the century. The vision statement summarizes the ambition:

> Tampere is a citizens' information society and a sustainable centre of expertise. Its operations are based on bold initiative, good public services, extensive networking and regional cooperation. (*Tampere News*, 2001)

On the basis of this strategy the municipal departments drafted their respective substrategies and outlined the development of their services. In addition, thematic strategies will focus on such issues as personnel policy, business development, housing policy and environment. The plan identifies a number of critical success factors, including 'utilization of the information society', 'innovation of service structures' and 'availability of a skilled workforce'. In this evaluation of e-Tampere we will try to establish to what extent these factors are being addressed.

Regional Context

Urbanization is a continuing trend in Finland and is quite visible in the Tampere region, Pirkanmaa, too. However, the natural beauty of the wider Tampere region has remained quite impressive. The e-Tampere programme for the information society is closely linked to the business strategy of the Tampere urban region, on the one side and the provincial development strategy on the other. The policy is to concentrate high-density planning in the urban parts of the region.

Over the last four years new partnership approaches have been established in order to enable the industries in Tampere and the Tampere region to take full advantage from the strengths of the region's high-tech enterprises and research institutes. The e-Tampere programme is strongly linked to the Tampere urban region's business strategy. Together with its neighbouring municipalities, Tampere is developing into an integrated metropolitan area. The size of the region and the way of life of the citizens seems to enable a balanced and sustainable development with an appropriate level of partnership between local government, citizens and enterprises. As the local people say: 'The city of Tampere is big enough and small enough at the same time'.

National Context

Tampere is one of the leading industrial cities in Finland. Several clusters of expertise are present in the Tampere region, not only in the field of information and communication technology, but also in such fields as mechanical and electro-technical engineering and biomedical technology. These sectors participate in the national Centre of Expertise programme.

Finland has a total population of approximately 5.17 million inhabitants. The origins of the Finnish GDP (the sum of all output produced by economic activities in the country) are mainly in services (62.5 per cent) and industry (33.7

per cent). Agriculture accounts for 3.8 per cent of GDP. The unemployment rate is at 9.79 per cent.[1]

The Finnish public sector accounts for 25.2 per cent of the total employment in Finland (1998). This is a consequence of a typical feature of the Nordic countries, where the basic social services are provided by the public sector. Local government employment accounts for 19 per cent of all employed labour force in Finland (Kasvio, 2001).

One of the consequences of the size of the public sector is that a rather substantial part of GDP is spent in the public domain. Local government expenditures alone add up to 17 per cent of Finland's GDP. The Finnish model does not, however, put too much of a tax burden on the enterprises. One of the unique features of the Finnish model for the information society lies in the way in which tax revenues are channelled into promising growth sectors. The Finnish National Technology Agency (Tekes) has financed virtually every successful technological enterprise in Finland, including Nokia. Furthermore, the Finnish National Fund for Research and Development (SITRA) provides seed money to incubators and firms at early stages of business development. This work is crucial to the development of Finland's information society programmes, including e-Tampere. This redistribution of tax revenues has become a key resource of the developing the information society, which has been identified as a 'survival project' for Finland (Castells and Himanen, 2002).

In recent times the size of the Finnish public sector has decreased rather significantly because of the privatization of certain *state* functions. At the same time, however, employment at the local government level has been steadily increasing. The local authorities in Finland have taken over some powers and responsibilities from the national government. These reforms enable Tampere to influence the well-being of its inhabitants more directly, e.g. by developing a stable and sustainable economy at the regional level. The prosperity of the region in the twenty-first century may largely depend on its own ability to develop an information-based industry. However, national and – increasingly – European policies will remain an important determining factor of the potential for economic growth and employment.

International Context

After two years of negotiation, Finland joined the European Union in 1995. The 1997 Tampere Strategic Programme explicitly added the dimension of international competition to the city's development strategy. It clearly stated the importance of representing the Tampere region in the decision-making bodies of the European Union. This strategy was certainly enhanced by the presence of a Finnish Commissioner for the Information Society (Erkki Liikanen) and

---

[1]   The Economist.com country briefings at htttp://www.economist.com/countries. Figures apply to year 2000. GDP breakdown: 1998.

the Finnish presidency of the European Union from July to December 1999, when Tampere hosted 11 EU meetings. The e-Europe programme (European Commission, 2000; European Council, 2000), launched in December 1999, is a model and one of the sources of inspiration for the e-Tampere programme.

By utilizing the existing knowledge base the ambition is to develop Tampere region into *a hub in the information society*. This asset is considered to be of prime importance for national and international partnerships. The city expects to become a preferred partner for national and international projects, facilitating the international links that are essential to industry, public services, education, research, art and culture. The second half of the 1990s saw a rapid development of mobile technology. For Finland and the Tampere region – the birthplace of mobile communications[2] – the development of mobile services has become a competitive advantage in the international arena.

Overview of the E-Tampere Programme

Tampere's vision and the strategy for the information society were put together in a comprehensive programme, called 'e-Tampere'. It ran from 2001 to 2005, with a targeted overall budget of €132m. In this programme the city of Tampere cooperated with private companies, partners from regional institutions of research and education and community organizations. The aim of these partnerships was to mobilize a wide variety of expertise and development resources. It was presented as a huge marketing effort to promote the city and the region. It was, in fact, 'branding' Tampere, all the way from the strategic planning level on to such practical operational matters as business cards. It is quite remarkable to note that almost every partner in the programme was able to present an e-Tampere business card, with pride.

The e-Tampere programme was prepared in the summer and autumn of 2000 by a programme planning team composed of representatives of the city of Tampere, the two universities, PCA Infocom OY, Prinnox Oy and the local branch of the Technical Research Centre Finland (VTT Automation). Most importantly, however, the members of the team were selected on the basis of their personal profile as experts in the field of ICT and organizing capacity. A large number of people participated in the design of the e-Tampere programme modules. Almost 50 key people were interviewed by the preparatory team. These preparations resulted in an e-Tampere programme with a prospected overall budget of €132m. Table 8.1 summarizes the available prospected funding for the e-Tampere programme.

---

[2] In 1974 Tampere saw the first NMT (Nordic Mobile Telephone) call in the world. It was followed by the first GSM (Global System for Mobile Communications) call in the world in 1991 and the first mobile Internet call in the world in 1995. Source: Tampere, Hermia Technology Centre Ltd.

Table 8.1   Funding resources for the e-Tampere programme

	2001	2002	2003	2004	2005	Total
City of Tampere	3,190,000	3,270,000	3,270,000	3,010,000	2,840,000	**15,580,000**
Unversity of Tampere	580,000	690,000	780,000	870,000	980,000	**3,900,000**
Tampere University of Technology	450,000	470,000	480,000	480,000	500,000	**2,380,000**
Academy of Finland	180,000	1,030,000	1,290,000	1,460,000	1,550,000	**5,510,000**
National Technology Agency	640,000	2,350,000	2,900,000	3,890,000	4,200,000	**13,980,000**
SITRA	150,000	150,000	150,000	150,000	150,000	**750,000**
EU/framework programmes	340,000	820,000	1,990,000	2,860,000	3,750,000	**9,760,000**
EU/ESF, etc.	30,000	30,000	160,000	190,000	280,000	**690,000**
Companies	790,000	1,620,000	3,070,000	5,350,000	7,180,000	**18,010,000**
Employment and Economic Development Centre	170,000	510,000	680,000	950,000	1,120,000	**3,430,000**
Engines/capital investments in companies	0	340,000	340,000	510,000	510,000	**1,700,000**
Accelerator/capital investments in companies	4,280,000	8,560,000	11,130,000	13,700,000	13,700,000	**51,370,000**
Technical Research Centre	260,000	260,000	260,000	270,000	340,000	**1,390,000**
Others	190,000	330,000	480,000	1,120,000	1,430,000	**3,550,000**
Total	**11,250,000**	**20,430,000**	**26,980,000**	**34,810,000**	**38,530,000**	**132,000,000**

The e-Tampere programme aims to strengthen the development of the information society by creating new knowledge, new businesses and new public services online. The ambition underlying the programme is that, by the end of the programming period, Tampere will be a global leader in the research, development and application of solutions related to the Information Society. It is an explicit choice to set high ambitions and demanding objectives, on the basis of which Tampere will attract people and businesses that are willing to commit themselves (Kostiainen, 2001).

The focus of the e-Tampere programme is on three main themes: strengthening the knowledge base of research and education; creating and accelerating the growth of business activities; and making public services available online (see Table 8.2).

Table 8.2   E-Tampere action lines, sub-programmes and budget

e-Tampere action line	e-Tampere sub-programmes	Budget (€) 2001–2005
Strengthening the knowledge base of research, education and training	ISI	17,160,000
	RELab	16,970,000
	EBRC	17,110,000
Creating and accelerating the growth of business activities	e-Accelerator	54,270,000
	Technology Engines	19,050,000
Making public services available online	InfoCity	5, 380,000
e-Tampere programme management	e-Tampere office	2,060,000
Total		**132,000,000**

In November 2001, on the basis of the e-Tampere approach to the information society, the European Commission awarded the city of Tampere the e-Government Label for one of the best applications of electronic government in Europe.

Programme Management

The e-Tampere programme is guided by a steering committee that meets about four times a year. It includes representatives from research and educational institutions, regional companies, public financers and the City of Tampere. The programme is executed by the so-called e-Tampere Office. This office has a small staff, including the director of the e-Tampere programme and an information officer. The e-Tampere office acts as a secretary for the steering committee and the working group. Furthermore, the office ensures communication and synergy between the relatively independent modules of the programme. It organizes events and meetings and is responsible for providing information and marketing.

Representatives of the key partners in the e-Tampere programme are gathered in the e-Tampere Working Group. This group of five to eight people supports the management of the programme in the practical coordination of the programme.

4   Strengthening the Knowledge Base of Research, Education and Training

The sub-programmes under this heading are the Information Society Institute (ISI), the Research and Evaluation Laboratory (RELab) and the *e*-Business Research Centre (eBRC). The prospected budget for these sub-programmes is €51.24m.

The Information Society Institute (ISI)

The Information Society Institute aims to promote the construction of an information society based on active citizenship on a local, European and global scale. The ambition is to make Tampere one of the world's leading centres of research and education related to information society by 2006.

In autumn 2001 the ISI was founded as an independent research centre affiliated with the departments of Social Sciences and Cultural Studies of the University of Tampere. The institute's mission is to promote and coordinate multidisciplinary research, development and educational activities with the participation of the University of Tampere and the Tampere University of Technology. The institute's activities are expected to contribute to the local needs central for the implementation of the e-Tampere programme. This is reflected by the key fields of research, which are:

- the user-friendliness of new information and communication technologies;
- the search of information and the use of services in electronic networks;
- the use of new ICT in teaching and learning;
- the use of new ICT in the provision of welfare services and in public administration;
- the social and economic dynamics of the information society;
- the communication cultures and structures in the information society;
- the development of innovation in business, leadership and work cultures;
- the longer-term perspectives and alternatives of developments in the information society.

The financial plan of the Information Society Institute amounts to €17.16 Euro, which is to be generated with the support of the two local universities, the City of Tampere, the Academy of Finland, the National Technology Agency (Tekes), the regional Employment and Economic Development Centre for Pirkanmaa, the European Union, companies and others.

The Research and Evaluation Laboratory (RELab)

The aim of the Research and Evaluation Laboratory is to test information society technologies, services and applications. The objectives are to introduce new ways of working that are of use to users and developers, to generate new businesses and to support new companies and new ways to do business in the ICT field.

RELab started in 2001 and is managed by the Tampere unit of the Technical Research Centre of Finland (VTT). The operations are divided into three parts that support each other:

* monitoring and researching new ideas and trends. Prospected results of this research include frameworks of reference to evaluate the viability of new products and services;
* research and development of next generation services in the first stages of their development cycle;
* evaluation and testing of products and applications that are in their final stages of development.

By evaluating, testing and developing the information society services, RELab plays a key role among in the e-Tampere programme. On the other hand, the city and the region of Tampere constitute an extensive base of user groups and serve as a test bed for new digital products and services.

The financial plan of the Research and Evaluation Laboratory adds up to €16.97m, which is to be generated with the support of the two local universities, the city of Tampere, The National Technology Agency (Tekes), the Technical Research Centre of Finland (VTT), the European Union, companies and others.

The E-Business Research Centre (eBRC)

The eBRC is a research centre specializing in practical and theoretical knowledge of electronic business and the new economy. It started in spring 2001 as a joint initiative of various departments of the Tampere University of Technology and the University of Tampere.

Research activities and the foundation of a network of professionals in the field of the new economy are the core business of the e-BRC.

The knowledge generated by the e-Business Research Centre will be utilized in the teaching and research work of the two local universities, as well as in the local business community. Thus it contributes to the sustainable development of Tampere and the Pirkanmaa region in the long run.

The financial plan of the e-Business Research Centre adds up to €17.11m, which is to be generated with the support of the two universities, the City of

Tampere, the Academy of Finland, the National Technology Agency (Tekes), the European Union, companies and others.

5   Creating and Accelerating the Growth of Business Activities

The sub-programmes under this heading are e-Accelerator and Technology Engine Programmes. The prospected budget for these sub-programmes is €73.32m.

E-Accelerator

The e-Accelerator programme aims to identify ICT-based companies and business ideas that have the potential to become a competitive player on the international market within two years. The programme will generate third-party venture capital for the selected firms. The ambition for 2006 is to have generated 20 to 25 top companies that will be listed on international stock exchanges.

The e-Accelerator programme started with a first competition for candidate firms in spring 2001. Altogether, 79 business plans were submitted, six of which were selected for a three-year development contract.

The e-Accelerator programme is based on a specialist network of Finnish business development and capital financing experts. The Tampere Technology Centre Ltd is responsible for running the programme and hosting the business incubator, with the support of Hermia Business Incubator.

The activities and the support of the e-Accelerator programme follow the stages of business development:

• marketing and training for starters (identifying potentials for the development of business plans to be submitted at the e-Accelerator competition);
• evaluation and selection of participating firms and supporting partners;
• pre-acceleration business development, planning and acquisition of capital;
• acceleration stage: product development, marketing, management development and training;
• global growth.

The e-Accelerator programme will contribute to employment and sustainable economic development in Tampere and the Pirkanmaa region. Five years from its inception, the 20–25 selected firms could employ about 1,500 people and have a combined turnover of €250m. The business development model and the infrastructure created by the programme may prove to be effective in the longer run too.

The financial plan of the e-Accelerator programme adds up to €53.27m, which is to be generated with the support of the two local universities, the City of Tampere, a number of established ICT firms and management supporting firms, venture capital, the Finnish national fund for research and development

(SITRA) and others. The total amount of third-party venture capital for the selected firms is estimated at €51.37m.

Technology Engine Programmes

The Technology Engine Programmes aim to produce an accumulation of knowledge on future technologies in the Tampere region. The focus is on information and communication technologies that will promote access and interaction. One of the ideas behind the Technology Engine Programmes is that easy access to Information Society services guides the technological developments in a more people-oriented direction.

The Technology Engine Programmes are coordinated by the Digital Media Institute at the Tampere University of Technology. They have been developed in cooperation with the University of Tampere, the Tampere Technology Centre Ltd and the National Technology Agency (Tekes).

The planning of the engine programmes and the consortia that execute the programmes were mainly established in 2001. The research activities started in the beginning of 2002 in five Technology Engine Programmes:

- adaptive software components (ensuring suitability for mobile communications and interactivity between independent software components);
- perception of information (understanding how human beings understand and 'process' information);
- broadband data transfer (developing advanced components, equipment and management tools for wired and wireless data transfer);
- user interfaces (developing natural ways of using electronic devices and services);
- neo reality (taking virtual reality one step further into value-added services and 'intelligent' spaces and equipment).

Each programme is supposed to include two or three projects. New products and businesses were expected to be generated by the beginning of 2004. The level of expertise in these research programmes should be among the world best by 2006.

The Technology Engine Programme ensures close cooperation between the various actors under the e-Tampere programme and stimulates an interdisciplinary approach. At the same time it integrates public domain considerations into the early stages of technological research.

The financial plan of the Technology Engine Programmes adds up to €19.050m, which is to be generated with the support of the two universities, the City of Tampere, the Academy of Finland, The National Technology Agency (Tekes), the European Union, the Employment and Economic Development Centre for Pirkanmaa, companies and others.

6 Making Public Services Available Online to All Residents (Infocity)

In the year 2000 62 per cent of the citizens of Tampere used the Internet. Of that number, 77 per cent visited the city of Tampere website.

The digital *content* offered by *local government* itself, mainly through www.tampere.fi includes city transport information (timetables, fares, routes), interactive library services (access to databases, reserving and extending the loan of books, personal profile of interest), map-linked information services (addresses, road works, zoning), regional organizations, online communities, online debates on municipal issues, digital proposals and their follow-up. The timetables for public transport are not only accessible via the Internet, but also via mobile communications on GSM and WAP.

The local authorities feel that *access* to the information society requires *infrastructure* and digital literacy. Free access to the Internet is to be available at the schools and via more than 100 terminals in public places such as libraries, hospitals, sports centres and municipal offices. These locations also offer training and assistance in the use of computers and the Internet. Finally, the Netti-Nysse offers access and training on wheels. This Internet bus stops at different business locations and residential areas on a regular basis.

The Infocity programme has a prospected budget of €5.38m. The Infocity project is aimed at developing Tampere's electronic government services. The objective is to make all services available online and used actively by all residents. The ambition is to make Tampere a model in the development towards the information society city, on an international rather than national level.

The Infocity project is coordinated by the Municipal Department of Communication (*not* the Information Technology Centre). Within the public administration five Internet posts were set up at the beginning of 2001, one for each sector. Furthermore, the project depends on cooperation between the administration, private operators and the third sector. Cooperation projects ranging from local district activities to national and international partnerships are underway.

Themes for the Infocity spearhead projects include electronic identification, electronic invoicing, online education, customer case management, electronic democracy and mobile services.

The Infocity programme addresses the *content* of the government services as well as the *access* to these services for the citizens:

* strengthening the resources and the service competence of the municipal Information Technology Centre;
* increasing the number of Internet content providers in the administration;
* providing more Internet training for city employees;
* developing the inhabitants' skills in using the Internet;
* increasing the number of public access Internet terminals in the city.

For the citizens of Tampere the Infocity project brings the e-Tampere strategy to life on a day-to-day basis, at the very beginning of the e-Tampere programme.

The development budget for the Infocity project adds up to €5.38m, at the expense of the city of Tampere. This budget is managed by the Department of Business Development. In addition, part of the regular budget of the Information Technology Centre will be spent on securing online services. The personnel costs incurred by the Infocity project are divided between the various city departments.

7 Conclusions

The proposals and sub-programmes of e-Tampere pay due attention to the required elements of content, infrastructure and access.

The e-Tampere strategy started from a mixed technological and societal angle: information society technologies. From the very start, however, business development was a secondary target, which is being equally well addressed. Almost by definition – but in reality too – the approach included the social and economic impact of the information society technologies from the outset.

The main competitors at the national level are Helsinki and Oulu. Helsinki is comparable to Tampere in many respects. The competitive advantage for Tampere will be mainly in terms of quality of life: large enough and small enough to be a pleasant place to work and to live. Oulu is extraordinarily good in mobile technology, but in the field of society and business development issues Tampere has a stronger profile.

Tampere's relative position at the international level has yet to be established. In fact, Tampere is a newcomer to the international platform, with a strong strategy. Some of the observations in this study may give a clue to possible directions.

There is a clear vision behind the e-Tampere programme. After its original conception in 1999 this vision has been elaborated in a process with a high level of participation of different actors. This process has increased the feasibility of the e-Tampere concept and consolidated a sufficient level of political and social support. Through this process of consultation, the vision was translated into a concrete strategy and the foundations were laid for the networks and the partnerships that are now in place to implement the strategy and to develop the projects.

E-Tampere has developed itself as a trademark. The PR is strong and very effective. This certainly helps to capture the opportunities and avert the threats. The effects of PR are mainly external. To maintain the support of local communities (business, citizens, politicians) PR has to be supplemented by active involvement. This involvement has been quite high at the drafting stages of the e-Tampere programme. However, once the programme had been drafted, it took some time to get the decisions implemented and to start up the sub-programmes. It was almost unavoidable that involvement of citizens and businesses was a little less at this stage. It appears that some actions should be taken to regain the same level of commitment and support as before. There are plenty of opportunities in the e-Tampere programme to accomplish this.

Strengthening the Knowledge Base of Research, Education and Training

This line of action is being executed quite successfully. The Information Society Institute, the Research and Evaluation Laboratory and the e-Business Research Centre are all in place and have started their work. The e-Tampere initiative and the City of Tampere have put considerable effort into convincing the research community of the potential benefits of cooperation. In the end the participating researchers are positive about the added value. However, now that these institutions are in place it may take some time to realize the potential. It may be easy to create visible results at the technological level but the changes in society that occur require a much deeper understanding. In balance, there is a need for patience and a proper monitoring of progress.

The business community has been very involved in the preparation of the e-Tampere programme. However, when the drafting of the programme was finished a period of 'radio silence' occurred. It took some time to get the programme approved by the city council and to start up the sub-programmes. Typically at this stage, the business community witnessed a kind of public bias in the process. At the time of writing, they feel that the research activities in e-Tampere are too dominated by public partners and they are somewhat reluctant to invest the proposed €20m in the programme. This is a serious matter, since in many cases co-funding from sources external to the region may depend on private sector involvement (e.g. for European Union R&D funding).

In general, however, the business community is very positive about e-Tampere. There appears to be solid ground for improved involvement.

The polytechnics and vocational training institutions are not yet actively involved in the e-Tampere programme. Their partnership was anticipated in the original plan, but concrete participation has not occurred until now. These institutions may be involved in the elaboration of the e-Education cluster.

In the beginning e-education received little attention. As there had already been a 'learning Tampere' initiative, there was no need to prioritize this topic in the e-Tampere programme. However, some partners, in particular in the business environment, felt that cooperation in the educational field could be much better.

The e-Tampere umbrella could help to develop an e-learning cluster. The ambition would be threefold:

- to study and to teach the impact of ICT on people (perception, human interface), business (e-business) and society at large (social impact). These issues are already being addressed in the programme in its present form;

- to implement e-education tools and methodologies in the educational and training systems in the Pirkanmaa region, from primary school onwards (knowledge base perspective, related to the ISI);
- to become a key player in e-education, in particular for meaningful pedagogical content production (knowledge base perspective and business development perspective).

Creating New Businesses Related to the Information Society and Accelerating the Growth of Existing Business Activities

This sub-programme appears to have made the most progress. The first round of selection for the e-Accelerator programme was a success. Three business plans out of 79 were selected for a three-year development plan. Furthermore, the Technology Engine Programmes also appear to be well in place. The level of cooperation between researchers in this line of action appears to be at a higher level than under that for the knowledge base (action line 1). This may be just a matter of stage of development of the lines of action, but it is too early to say.

The focus in the line of action for business development is a little biased towards technology-related businesses. It covers mainly hardware and embedded software. There is hardly any appearance – if any at all – of business development in services and applications. Such an angle would extend the scope of the e-Tampere project from e-business to any business, making it more relevant not only to the wider business community but also to citizens and public institutions.

It would be advisable to add product and market development for services and applications to the e-Tampere strategy, perhaps with particular focus on:

- the public sector, including, for example, municipal services, healthcare and education;
- small and medium-sized enterprises.

The development of business schools is related to this issue. The original idea (December 2000) was to establish an e-Economy Business School for training and research activities. From the very beginning, however, the e-Tampere integrated the training components as standard elements in the graduate, postgraduate and executive education programmes of the Tampere University of Technology and the University of Tampere. So far, this has not given sufficient profile to the business school(s) of Tampere. An approach more like the ISI and the TEP (which have been set up in two slightly different ways) seems to be needed.

The curriculum of the business school would include, amongst others things:

- productivity gains through the use of ICT;
- product and market innovation with ICT;
- new methods of working and organizational design.

Market development also requires a cohesive approach to testbeds. After the stages covered by RELab, larger testbeds are required. Such large testbeds (a neighbourhood, a district, or perhaps the whole city and region) would probably require a broadband infrastructure and applications with a large consumer base, such as public services. A first consequence would be that infrastructure development and the development of public services would have to be more related to the strategy for business development. A second consequence would be that the concept of testbeds for applications and service development would constitute 'the missing link' between fostering knowledge, promoting business and serving the public interest.

Making Public Services Available Online to All Residents

Somehow the e-Tampere programme seems to undervalue the importance of services in general and public services in particular. This may be because the programme in itself represents a shift of focus: away from the emphasis on applications in the dot.com hype, towards enabling technologies (engines) and business development.

The scope of the agenda to 'make public service available online' is a little narrow. It does not seem to have the same priority as the other two projects (the knowledge base and business development). Some observations illustrate the difference.

- External partners are not involved in the design and the implementation of this action line and the Infocity project. There is no evidence of public-private partnerships in this field. Private partners tend to be more attracted by the other lines of action (in particular technological knowledge and business development). On the other hand, local government does not seem to have positioned itself as an interesting partner, a change agent or a leading consumer. Perhaps there is a need for a clear vision of the role and position of ICT facilities in public service delivery, in particular from the political and the managerial point of view.
- So far, the budget for Infocity seems to be have been made available by the municipality only. Unlike in other projects, there is no prospected co-funding by firms, national agencies and the European Union. This in particular illustrates a lack of ambition, marketing effort and perhaps self-esteem. In the other projects the partners involved will generate an average of 93.5 per cent of the budget.
- As a consequence the budget for this action line is very limited. Although it accounts for almost 35 per cent of the municipal input, it accounts for less than 4 per cent of the total e-Tampere financial plan. This is quite remarkable when compared to the importance of the public sector in the Finnish national system. The public sector accounts for approximately 25 per cent of the

employment in Finland and government spending accounts for 17 per cent of the Finnish GDP.

- The ICT department does not appear to be heavily involved in the e-Tampere project. It is not even in charge of the Infocity project.

The difference in weight attributed to this line of action may have its cause in a paradox: Tampere was doing quite well in electronic public service delivery at the time the e-Tampere programme was designed. It was probably the most advanced city in Finland and no extra efforts were required to keep that position. Consequently there was no sense of urgency. However, as has recently been shown at the e-government events in Brussels, the pace of development in this field is rapid. An extra effort is needed for Tampere to play in the world league and to remain in the top of best practices in Europe. Spin-offs from the other action lines (knowledge and business development) can certainly be of help here, but the public services line of action also needs a profile of its own. The performance of local government itself is often seen as an indicator for the potential of development in the social and economic domain.

References

Breslau, K. and Heron, K. (2000), 'The Debriefing: Bill Clinton', *Wired Magazine*, December.

Castells, M. and Himanen, P. (2002), *The Information Society and the Welfare State – The Finnish Model*, Oxford University Press, Oxford.

City of Tampere (1997), *Information is the Key to the Future. Tampere's Municipal Policy for the Twenty-first Century.*

City of Tampere (2000), *eTampere Programme Plan (2000)*.

City of Tampere (2001), *Kaikem paree Tampere [Tampere, the Best of All]*.

Digital Media Institute (2000), *Annual Report 1999*, Tampere University of Technology.

European Commission (2000), *e-Europe – An Information Society for All*, EC, Brussels.

European Council (2000), *Presidency Conclusions*, Lisbon, 23–24 March.

The Economist (2000), 'Government and the Internet: Haves and Have-nots', 24 June.

HERMIA (2001), *Annual Report 2000*, Tampere Technology Centre.

Janssen, D., Kampen, J.K., Rotthier, S. and Snijkers, K. (eds) (2003), *De praktijk van eGovernment in zeven landen van de OECD*, Steunpunt Bestuurlijke Organisatie Vlaanderen, Leuven.

Karvonen, Erkki (ed.) (2001), *Informational Societies. Understanding the Third Industrial Revolution*, Tampere University Press, Tampere.

Kasvio, A.J. (2001), *Towards a Wireless Information Society: The Case of Finland*, material for a lecture series given in Autumn 2001 at the Univesity of Tampere: http://www.info.uta.fi/winsoc/engl/lect/progr.html.

Kasvio, A.J., Laitalainen, V., Salonen, H. and Mero, P. (eds) (2001), *People, Cities and the New Information Economy*, Palmenia-kustannus, Helsinki.

Kostiainen, J. (2001), 'eTampere – Generating Growth through collaboration', in Kasvio, A.J., Laitalainen, V., Salonen, H. and Mero, P. (eds) (2001), *People, Cities and the New Information Economy*, Palmenia-kustannus, Helsinki.

Kostiainen, J. and Sotarauta, M. (2002), *Finnish City Reinvented. Tampere's Path from Industrial to Knowledge Economy*, MIT Industrial Performance Center, Cambridge MA.

Tampere News (2001), 'Tampere's Strategy Programme Completed', 2 November.

UNDR (2001), *Human Development Report 2001: Making New Technologies Work for Human Development*, United Nations, Geneva.

Virtanen, M.K. (2002), *eTampere, Aiming at a ModelCity of the Information Society. Reports, Comparisons and Citizens' Assessment*, Information Society Institute, Tampere.

Interview Partners

Harri Airaksinen, Director, Business Development Centre, City of Tampere.

Hannu Eskola, Director and Professor, Digital Media Institute, Tampere University of Technology.

Harri Jaskari, Managing Director, Tampere Chamber of Commerce and Industry.

Antti Kasvio, Research Director, Information Society Institute, University of Tampere.

Juha Kostiainen, Director, Business Development and Marketing, YIT Construction, Helsinki; former Director of Business Development for Tampere.

Jarkko Lumio, Managing Director, Oy Media Tampere Ltd.

Kari Löytty, Business Development Officer, Business Development Centre, City of Tampere.

Pekka Markkula, Development Manager, Information Society Unit, Sonera Corporation, Tampere.

Heikki Pettilä, senior citizen, ATK Seniorit Mukanetti ry, Tampere.

Jari Seppälä, Head of Information, Tampere City Hall.

Jorma Sipilä, Rector, University of Tampere.

Jouko Suokas, Research Director, VTT Automation, Tampere.

Mika Uusi-Pietilä, Director, Knowledge Creation-IT, Minutor Oy, Tampere.

Lea Vakkari, Project Director, Hermia Tampere Technology Centre.

Toni and Jani Vakkari, junior citizens, schoolchildren.

Markku Valtonen, Senior Delegate, Tampere Central Region EU Office, Brussels.

Jarmo Viteli, Executive Director, eTampere Office.

Chapter 9

The Case of The Hague

Introduction

The city of The Hague is very ambitious in the field of ICT policy. In this case study, we will analyse The Hague's electronic governance using the framework developed. We focus on two major projects: Residentie.Net and Transparent City Hall. Furthermore, we pay generous attention to the way The Hague presents itself on the web – in the words of our framework, the city's content organization. We start in section 2 with a brief history of The Hague's ICT policy. Section 3 contains a description and analysis of the city's e-government policy. Section 4 focuses on the Residentie.net project, an award-winning project to create a virtual digital community in the city. In section 5 we move to the subject of content organization: we critically assess the city's presence and presentation on the Internet. Section 6 is about computer- and Internet-access policies and in section 7 we discuss electronic infrastructure policies and initiatives. Section 8 concludes and provides policy recommendations for improved strategies.

2   Urban E-Vision

This section introduces The Hague's ICT policy efforts. We first briefly sketch the history of The Hague's ICT policy. Second, the latest ICT strategies are summarized.

History

In the 1980s The Hague's ICT policies were restricted to office automation projects. It was not until the 1990s that the awareness that ICT would have major consequences gained ground, not only for work practice within the municipality but also for networks relations with citizens and other urban actors. The breakthrough of the Internet played an important role in this process: it opened a new perspective of a 'wired society', in which people communicate and interact in radically new and hitherto unexpected ways. The wave of publications on this new paradigm urged many urban policymakers to rethink their strategies in the light of the new challenges and threads of the information society. The same happened in The Hague. In the early 1990s, Internet activity in the city of The Hague was very low. Like in other cities, a few 'freaks' used and experimented with the Internet and that was it. At that time, pioneers united and created the 'virtual city of The Hague'. Over the years this became an important community of advanced Internet users.

Meanwhile the municipality had come to realize the importance of the web. Halfway through the 1990s, the City of The Hague did not have an official website. Then it was discovered that The Hague was on the web: an enthusiastic individual civil officer had apparently published a municipal website on his own initiative, with some basic information on the city and the municipality.

From the second half of the 1990s onwards developments in the field of e-government moved very quickly. At this stage of development (1996–1999) a special Urban Telematics Project Office was responsible for raising awareness and developing pilot projects. This office provided the institutional driver for telematics within the administration and in the city. As in many other cities, this office enabled the acquisition of new skills and attitudes, a break with the bureaucratic traditions of public administration and the formation of partnerships with the private sector. Furthermore, through the Telecities network, the city actively looked for an exchange of experience and partnerships with other European cities.

The city initiated and coordinated the European Commission-funded 'Infocity' project, which was a catalyst to bring municipal information online: later in the 1990s this project was continued with the support of the VirtueHalls project. At the time of writing, the city is heading towards being in a position to offer all its services on the web. ICT is at the heart of The Hague strategy. The mayor and the alderman for information policy strongly promote and support The Hague's activities to create a transparent city hall and digitalization of municipal services: they also initiated policies to promote access for all citizens of The Hague. In 1999 the city produced an e-vision, in which the city's ambitions for the next few years were laid out.

Overall Vision, E-Strategy and Key Projects

The Hague's latest (1999) e-vision is built on three ambitions:

1   empoweringing citizens for the digital age;
2   transforming the municipality in a demand-led, transparent and flexible organization;
3   strengthening The Hague's 'urban identity' as an international city of peace and justice and a city with a high quality of life, culture and tourism, among other things with the help of ICTs.

The ambitions are translated in seven strategic projects:

1   Residentie.net. This is a platform for citizens where they can find information, organize themselves on the web. The project also includes a free Internet

provider for all The Hague citizens and enables Internet access not only via the computer but also using telephone and TV.

2 The transparent city hall. This is The Hague's e-government project, involving the demand-oriented digitalization of municipal information provision, improvement of service delivery and the application of technology to improve decision-making processes.

3 Learning in the information society. This project involves several actions to empower and encourage citizens to use new technologies and to use technologies for new forms of education.

4 The Hague online. This programme will strengthen The Hague's identity as an international city of peace and justice, and to support democracy with online facilities.

5 The Hague Telecom City. The Hague has the ambition to host a concentration of telecom-providers. Within this project, the city seeks to strengthen its position as an attractive location for this branch of the ICT sector.

6 Telematics in transport, a programme that focuses on mobility and accessibility of the city.

7 Information Workplace, a think-tank to encourage bottom-up initiatives in the city. Its mission is 'to develop a development model for the city in the information society, together with the city' (Den Haag, 2001b, p. 3). One of the key roles of this organization is to develop or promote the application of ICTs for the benefit of the city and its citizens in many fields, for example, in health care, learning, safety, culture etc. It is a 'practical think-tank' that develops visions and strategies and translates these into pilot projects in the city.

Not all the ambitions have been realized, and some projects have been cancelled or changed in the years after 1999. As we will see later in this chapter, the Residentienet project has undergone important modifications, and the free email service has not materialized. The 'information workplace' is now an independent organization with no formal links to the municipality.

In 2003 a new ambition was formulated: to promote the provision of broadband to individual households in the city. At the time of writing, feasibility studies are being carried out to find out how to connect every citizen to fibre-optic infrastructure, and how to involve the current telecom players in this initiative.

3 E-Government Initiatives

One of The Hague's key ambitions is to become a responsive and demand-led organization. To this end, it started the Transparent City Hall initiative. We start this section with a description of the Dutch national context. We will discuss the role of the national government in the promotion of urban e-government. Next, we will outline the architecture of the Transparent City Hall initiative. Note that

Table 9.1   Milestones in The Hague's ICT policy

1994	Creation of two Virtual Cities: Digitale Hofstad and Digitaal The Hague; No involvement of municipality
1995	Creation of municipal Internet site by enthusiastic civil officer
1996	Creation of the Urban Telematics Project Office
1997	First telematic centre opened in public library
1998	Start of explicit urban ICT policy: ICT chapter in City Council's Strategic Plan 1998–2002
1999	Publishing of 'Urban E-vision'
1999	Start of the Transparent City Hall project, aimed to bring municipal services online, improve citizens' commitment to local government and improve participation. Budget: €2.7m
1999	Start of Residentienet, a digital platform for citizens of The Hague and provision of free Internet access for all citizens
2001	The Hague becomes one of the three Dutch 'superpilot cities' to work out solutions for e-government policies
2001	The Hague announces strategic partnership with Microsoft for e-government
2003	Expected completion of Transparent City Hall project

this is only a plan that recently has been developed. It is too early yet to make a judgement on the actual execution of it. Lastly in this section, we take a more analytical stance.

National E-Government Policies

The Dutch government has the information society development high on its agenda. Within this broad theme, e-government is a key subject that receives generous attention. The coordination of e-government initiatives is in the hands of the Ministery of the Interior.

This ministry has set up a dedicated ICT-implementation organization for the public sector, ICTU. Many programmes run under this umbrella, some of which are particularly relevant for our purposes. A key programme deals with electronic identification solutions and electronic security issues. At the time of writing the 'public key infrastructure' (PKI) is being tested in a laboratory situation. The first pilot projects were launched by the end of 2002 (PKI newsletter September 2001; www.pkioverheid.nl/informatie/experimenten.htm). Digital identity is put onto a smart card that enables transactions with public agencies. One of the reasons for the development of this PKI was to have a single, reliable identification infrastructure and avoid the development of many different identification schemes. However, several public organizations did not want to wait for this national security system and had already introduced other types of identification and signature. An example is the Dutch Tax Service which at the time of writing already allows transmission of very sensitive tax information through the Internet using a digital signature system.

A second key national project is named Public Counter 2000 (OL2000). It aims to establish demand-oriented 'one-stop shops' at all kinds of government organizations, to stop fragmentation of the services offered to citizens. It also plays a role in helping cities to introduce online services. Among many other things, a list of online municipal products and services (a catalogue) has been drawn up. The organization has developed practical instruments to set up online services and help public agencies that want to introduce them. The website also contains a list of companies with expertise in implementing e-government solutions.

Third, national government puts a high priority on the dissemination of good practices of e-government. For instance, a centre of expertise was set up as a platform for knowledge and experience exchange on innovative decision-making. Also in the OL2000 project much attention is paid to the practical implementation of e-government solutions. On the Internet, all kinds of user guides and good practices are downloadable, for instance on 'how to create a single counter', or 'how to develop pro-active services' (see www.ol2000.nl).

A third relevant action of national government is the appointment of three 'superpilot regions'. These areas should build a lead in electronic services delivery, and serve as examples for other cities and towns in the country. The three superpilot cities are The Hague, Eindhoven/Helmond, and Enschede. Each of the regions obtains a subsidy of €2.7m.

E-government in The Hague: The Transparent City Hall Project

The City of The Hague has incorporated the superpilot projects as integral part of its e-government strategy.

The envisioned architecture has a number of features. It starts by identifying a number of points of interaction between citizen and municipality. These can be physical or virtual desks. The next step is to connect the desks to the municipal contact centre. This centre enables any citizen to access demand-oriented information on a 24/7 basis. At physical or telephone desks, staff can access the site for the client. Every client has a certain profile. According to this profile, a client has access to municipal services. In the architecture, every client profile matches with a catalogue of municipal 'products' (allowances, documents, etc.).[1]

Next, a link should be made between the municipal products and the legal regulations they are based on. This link is important, for instance, to establish whether an individual client has the right to access certain data or is eligible for financial support, etc. The system also enables citizens to be alerted when the situation has changed, for instance when legislation changes, when important

---

[1] This list is designed by the Ministry of the Interior.

decisions are made by the city council or when an individual becomes unemployed. All these changes affect their use of and right to municipal services.

After it has become clear that a citizen has the right to access, to buy a range of municipal products or is eligible for an allowance, a transaction has to take place. For every municipal product, a digital form is created for the transaction. Every transaction has its own process – these processes are to be standardized. Some transactions are complicated, for instance, if you want to start a restaurant in the city, you need a lot of different permits. Every transaction becomes standardized and is translated in actions for the individual municipal departments that have to do the job. Importantly, the individual steps of the process rules are stored in a databank that is accessible for the client. Thus, at any moment a client should be able to check the status of the process for his or her transaction (comparable to the tracking and tracing systems of express mail couriers).

Transparency means that the entire process of municipal service provision should be visible and traceable. This implies that every citizen or company that buys or receives a municipal service should be able to check the 'municipal internal logistics' at any moment in time, comparable to the tracking and tracing facilities that mail express services offer. This has consequences for both the external (i.e. the interface with the clients (firms and citizens)) and the internal organization.

The Hague has opted for an organizational model in which the internal departments of the municipality remain intact. A 'virtual organizational layer' is being created to increase transparency and demand-orientation. The city works with client profiles.

The transition to a more transparent model involves substantial investment. The project had a budget totalling €2.9m for the period 2000–2003 (Den Haag, 2001a). The Dutch national government provides €450,000. It has appointed The Hague as one of the three 'superpilots' in the field of municipal e-government.

Property tax
The overall Transparent City Hall project should be the umbrella for all e-government initiatives of the city. However, this is not the case. The most remarkable e-government-project in The Hague is the 'WOZ infodesk' service, which is not part of the Transparent City Hall initiative. It works as follows. In The Hague – as in all Dutch cities – the city raises local property taxes, based on the value of the real estate objects. The WOZ-infodesk website allows owners and users of real estate (citizens and companies) to check the value of their objects,[2] obtain the taxation reports and eventually respond electronically. Access requires a username and a password. It is a very practical service and widely used and known by the general public.

---

[2]   They can also check a limited number of similar objects, to enable comparison.

Analysis of E-Government

The Hague's overall e-government concept suggests putting the client (citizen or company) in the driver's seat. This is a good starting point, and breaks with a supply-oriented tradition of public service delivery. The fact that the city was appointed a 'national superpilot' is an indication of the city's advanced standing in this respect. The city has not deliberately opted for a radical organizational redesign. The explicit choice of a 'virtual layer' instead of a new organizational structure may help to implement changes in the organization that will very probably add to the quality of municipal services. However, in the longer run, current structures may prove inefficient as means to deliver fully-fledged and integrated services to citizens and companies. Building the new information system on old structures may be a burden for future innovation. High levels of flexibility will be needed.

The Hague's e-government project suggests that e-government brings about new relations with non-municipal partners: it is a catalyst for all kinds of new network relationships. From this perspective, as part of e-government pilots the city planned to improve the service delivery for people who want to get married. This would be done by introducing electronic handling of the formalities for couples. Furthermore, a portal would be designed to show all the localities in the city where the marriage can take place. It would allow for online reservation as well. In this project, the traditional role of municipality would be digitalized and probably made more efficient and client-friendly. In the longer run, private businesses will be involved as well. It is possible to create a one-stop virtual shop that offers all marriage-related services, not only the official municipal services but also catering, flowers, restauration, logistics, etc. That would entail a new role for the municipality as information broker as well as service provider. In a more scaled-down option, the municipality could sell the information it has – on marriage date, addresses of the couples – to all kinds of business partners who might be interested in offering the couple a range of products and services. Evidently this would require the full permission of those involved. However, in the course of the project, many difficulties arose. In practice, the city did not manage to involve all the actors and gain their cooperation in the project.

The involvement of technology partners

A key issue when introducing e-government is the involvement of technology partners. In The Hague, as in many other cities, many information systems work alongside one another.

The municipal organization is split up in 14 main directorates. All of them have their own information technology departments. They operate the main information systems applications as well as the computers, the Local Area Networks and the mid-range computer systems. After the outsourcing of the municipal computer centre in the early 1990s, maintenance of the technology and the information systems, as well as mainframe operation, were contracted out to a preferred supplier for a period of five years. After this outsourcing contract expired these tasks were contracted out by the individual departments in a rather uncoordinated way. There is no information systems integration. The telecommunications infrastructure, however (office cabling, metropolitan infrastructure, data communications, mobile and fixed telephony), is highly integrated in a central unit called Haagnet, which was created in the early 1990s. The actual operation of all communication facilities is contracted out by Haagnet.

For e-government to be effective and efficient, system integration is an important precondition. The city was searching for a partner to integrate its various information systems in a comprehensive way. Here Microsoft enters the picture. In early 2001, the city announced a 'strategic partnership' with Microsoft to connect the systems to create the envisioned integrated e-government solutions. This announcement evoked heated debates in the press – local and national – and in the city council about the role of Microsoft. There were several main objections against Microsoft's involvement. First, many commentators were worried that involving Microsoft would imply that the e-government services would only be accessible using Microsoft's Internet browser. Second, it was feared that Microsoft would use proprietary systems and file formats instead of open standards. This might result in a lock-in situation where only Microsoft had the key to run the system and to open files, which would give Microsoft too much power. The company does not have a reputation of being very open. Microsoft adversaries suggested using Linux open source system instead, which would guarantee access, but the problem with Linux is the lack of support for its operations.

We conclude that much of the fuss was a little premature: the responsible aldermen had only expressed an intention to engage in a strategic partnership with Microsoft. The exact conditions had not been set. Many of the difficulties could have been avoided with a more careful presentation and communication. The high-level joint press conference of Microsoft and the City of The Hague in particular raised much concern. This case therefore shows the importance of thoughtful communication to the public at large when 'sensitive' companies like Microsoft are involved.

At the time of writing, it is still unclear what Microsoft's exact role will be. It has developed the .Net technology that enables communication between many types of systems and file formats. Nevertheless, economic theory suggests that lock-in is a real threat in vital ICT systems (Shapiro and Varian, 1998). One of the advantages of having Microsoft involved is the potential to break the existing 'duopoly' in The Netherlands, where two companies (Roccade and Centric) play a very dominant role in the municipal information systems. If Microsoft develops as a third key player, competition in this market will increase, with possibly lower prices and more innovation.

The case of The Hague shows that introducing e-government is very difficult. In 1999, high ambitions were formulated in the action programme of the city. However, at the time of writing results are meagre. Only some forms have been put online and there is a preparation of the future 'information architecture' that should allow for more transparency. It will be years before this architecture becomes reality. Unlike in several Scandinavian and German cities,[3] no online transactions are possible, basically because there is no system for online identification. The slowness of the process cannot be blamed on a lack of political will: the mayor and the responsible alderman strongly promote The Hague's informatization strategy. Rather, we gained the impression that on the level of individual departments, there is substantial resistance to changing work practices or even sharing information. An internal communication strategy is needed to convince every department involved. Furthermore, incentives should be invented to move them in the right direction.

The transparent city hall is in the making, but the city already boasts one e-government success: the WOZ-infodesk. Curiously, the development of this web is not linked to the 'Transparent City Hall' project. Rather, it is a 'stand alone' initiative of the local tax department of the city. It rests on the cadastral data and data of the municipal tax department. This example shows that a successful e-government application can be more easily obtained when only one department is involved.

4 Local Virtual Community Formation

This section deals with the second major ICT project in the city: the Residentie. net. This project aims to stimulate citizens to use Internet as well as to create web content and form local digital communities. We describe the project in the first and then analyse it.

Residentie.net

The project offers free[4] Internet access for all citizens, including those without a computer. Both Internet-based technology and interactive videotext are used to provide free Internet access and e-mail to every citizen and company in The Hague. Furthermore, the website Residentie.net is meant as a platform for the creation of digital communities. Citizens can create thematic 'squares', for instance, a square for your own physical neighbourhood. The site provides very simple and user-friendly tools to create these squares, but also to make homepages and participate in newsgroups. The website offers local information on events, etc., as well as services. It also contains a virtual marketplace for second-hand goods. Finally, the site also aims to serve as platform for discussions on issues that concern the

---

[3]   Bremen (Germany) uses digital signatures for online identification. Users need a card and a card reader to identify themselves. See www.bremer-online-service.de.
[4]   There is no monthly charge. Users only pay for the connection with the ISP server.

city, such as city developments plans. Residentie.net is a joint initiative of the municipality and two telecom network operators, KPN (the incumbent telecom operator in The Netherlands) and Casema (a cable company), and was financially supported by the TEN-telecom programme of the EU. KPN is responsible for the technical part of the website and its commercial exploitation, while Casema is responsible for the interactive teletext version. The Residentie.net foundation employs five people to manage the content of the website. The municipality initially invested €1.13m in the setting up of the project. The exploitation costs – carried by KPN – are around €730,000 annually. The idea was that the project would be profitable after three years, but this target proved hard to reach. At the time of writing, KPN is considering withdrawing from the project. Because of its solvency problems it has to cut unprofitable business; furthermore, the company no longer regards content creation and management as part of its core strategy.

Analysis

The ambitions of the project were sky high and daring. By that time it was the only local initiative in Europe to offer tools for every citizen to create web content. The concept combines access, content, community formation and e-democracy in an innovative way. Table 9.2 shows that the Residentie.net already counts a substantial number of digital squares. We found high numbers of squares in the subject fields of sports, 'my city' (all kinds of urban affairs), government/politics and entertainment. If you look at the activity level of the squares, the same categories are strong, but also 'games and computers' are high in the list.

In our view, the number of visitors of Residentie.net is relatively low. The site attracts around 25,000 visitors monthly: 90 per cent access the site via the Internet and 10 per cent use the teletext services. This makes it one of the busiest sites in The Hague but we consider that the figure is modest given the ambition of The Hague to have 'all citizens connected' and to make Residentie.net *the* urban platform for information and communication. Internet penetration in the city has passed the 60 per cent mark. Thus, in a population of 443,000 there are 265,800 frequent Internet users. This implies that only 9 per cent of these users visit the Residentie.net.

A close examination of the website reveals that the active and passive presence of local companies and organizations is quite low. First, only few local companies can be found on Residentie.net and even fewer are ready for e-commerce. The website claims to be a portal for virtual shoppers in The Hague, but in fact the site does little more than listing some addresses of individual 'bricks and mortar' stores. This list is very restricted. Second, there is very little activity of other types of local organizations – sports clubs, churches, voluntary organizations, community organizations, etc. – as well. Evidently these actors have little interest in participating in Residentie.net, or do not even know of its existence. At the time of writing, Residentie.net had insufficient 'mass' to be considered as a virtual

Table 9.2 Virtual squares on Residentie.net

Square categories	Total number of squares	Total number of postings	Average number of postings per square
Sports	119	770	6.,47
'My City'	99	811	8.19
Remaining squares	96	398	4.15
Inactive squares	92	230	2.50
Government and politics	50	238	4.76
Entertainment	44	328	7.45
Going out and culture	40	211	5.28
Games and computers	37	347	9.38
Hobbies and animals	31	204	6.58
Movies and music	29	100	3.45
Family and friends	27	202	7.48
Healthcare and well-being	25	72	2.88
Work and education	24	100	4.17
Residentie.net	21	149	7.10
Citizens of The Hague	18	129	7.17
Food and beverages	12	72	6.00
Religious and spiritual	2	2	1.00
Total	766	4363	5.70

The Hague. Our interviews suggest that advertisement and sponsoring revenues are low. This can partly be explained by a weak online advertisement market, but interviews suggest that the score could be higher if more efforts were made. During its short life the website has been plagued by technical problems. Half way through 2001, for instance, the site proved almost impossible to access using Netscape Communicator.

Residentie.net is a place for *spontaneous* discussions among citizens. Table 9.3 gives an insight into the number and subjects of the online discussions that take place on the site. For a city of this size, we think the number of discussions is rather low.

Finally, the Residentie.net is the vehicle for organized discussions. In 2001, on the website, the municipality organized an online discussion about the future design of a park in the city. The number of participants was 'unexpectedly high' although none of the civil officers we interviewed could give us exact data. Interestingly, not only citizens participated in the discussion, but also other users of the park, such as employees of companies nearby. This would never have happened if the discussion had been organized in the traditional way.

In December 2001, three discussions were held, one on youth policy, one on a redevelopment plan for a main street in the city and one on street violence.

Table 9.3   Discussions on Residentie.net, state-of-the-art, 4 January 2002

Topic	Number of discussions	Most recent start date	Total number of postings
City affairs	6	04 December 2001	15
Culture	1	27 November 2001	2
Sports	5	14 November 2001	19
Music	2	20 October 2001	19
Politics	11	01 January 2002	20
Internet	2	11 November 2001	4
Calls	10	03 January 2002	20

For each of the discussions, campaigns were held to involve as many people as possible. Also, politicians stressed that the outcomes of the debate would be seriously considered. Nevertheless, it proved very difficult to engage many people: the most successful discussion – on youth policy – drew only 30 people into the chat room.

The business model
The business model of Residentie.net has not worked as expected. One of the key problems is the role of the main partner, KPN. The exploitation of Residentie.net was put in the hands of KPN. This company was interested in the project for three reasons. First, KPN considered the project as a way to acquire 400,000 new clients – every citizen of The Hague online! – with subsequent positive impact on the value of KPN stocks. Second, part of KPNs international strategy was to provide not only infrastructure and connections, but also to move into content. It is important to realize that by that time (1999 and early 2000) the company was in the middle of the dotcom boom. Expectations on returns from online advertising were very high and payback period mattered less than rapid expansion. A few years later, KPN's engagement has not worked out well. In the first place, one may wonder whether a technology-oriented company is good at managing content at all. But perhaps more importantly, the resources put into the project were insufficient to meet its high ambitions. It is very hard to fill the site with news, engage all local firms and organizations, help setting up squares and manage discussions with an editorial board of five people. The claim of some that KPN has been slow to approach companies and other organizations in the city to participate or to advertise on the website should be seen in this context.

Next, there are complaints about the technical set-up of the site. One issue is the quality of the interactive tools on the website. On the one hand, experienced computer users complain that the tools are insufficient and inflexible. On the other, the tools are rather complicated for absolute beginners: sport clubs and voluntary organizations would very much welcome more guidance and support when setting up their website. Second, there have been technical problems concerning accessibility and quality of the website. Within KPN, the creation of content and the technical maintenance of the web take place at two separate locations, namely The Hague – the site where the editors have their base – and Utrecht. This has caused many misunderstandings. Internal communication problems at KPN were a key reason for technical problems in the recent past. Initially, KPN used its off-the-shelf Internet technology named 'het net' which it already had in use. This format proved unsatisfactory for an interactive community-like website, and was changed later on.

One of the key reasons for the low 'mass' of Residentie.net lies in the absence of links with other ICT-related initiatives in the city. The clearest example is the lack of integration with the municipal library's ICT access policy. In The Hague, the library offers very cheap courses for local organizations that want to develop a website. It would seem natural to offer the applicants the possibility of becoming active on Residentie.net: this could strengthen Residentie.net's base. But no such thing happened. Second, in the set-up of Residentie.net, there has been no cooperation with the existing 'digital city'. The Hague already had a very active digital community who were to a large extent bypassed.

We conclude that the City of The Hague put too much responsibility in the hands of its technology partner who, in the end, showed insufficient commitment and/or expertise to do the job. The project was 'thrown over the fence' and probably over the wrong fence. Several interviews suggest a lack of interest of policymakers – politicians and civil officers – in the well-being and potential of Residentie.net. It is illustrative that, for the organization of online debates, the Residentie.net editorial board had great difficulties in getting people in City Hall involved.

5   Content Organization: The Hague on the Web

The next topic – closely linked to the Residentie.net discussion – concerns the way The Hague is represented on the web. The city has opted for a separate website for municipal affairs. How is digital content organized in The Hague, and who organizes which content? Table 9.4 summarizes the available websites. We will briefly discuss them below.

Www.denhaag.nl

This is the city's municipal website. The information is organized for inhabitants, companies and visitors. The site has several interactive features. It permits downloading forms for a wide range of municipal services and it allows for complaints about municipal services provision or comments on policy. The site does not yet allow access to services for which identification is required. It enables access to a database of city council decisions and policy documents of the last 10 years. Online payment is still not possible. The site is maintained by civil officers.

Table 9.4   The Hague websites

Website	Target groups/ content	Management
www.denhaag.nl	Municipal information and services for citizens, companies and visitors	Municipality
www.thehague.nl	Foreign companies	Municipality
www.denhaag.com	Tourist information	The Hague Visitors and Convention Bureau
www.Residentie.net	Local information + communities	Residentienet

Www.thehague.nl

This site offers information for companies that might be interested in locating in the city of The Hague. The site looks good, but a major weakness is the absence of links to other websites. Furthermore, the site only presents very generic information and offers few interactive possibilities. Basically, the site only consists of an 'address book' showing visiting addresses of embassies, religious denominations, international schools and clubs. Under the heading of 'Education facilities', the universities of Leiden and Delft are mentioned: however, there is no mention of the Haagse Hogeschool. As an inward investment website this site is badly flawed.

Www.denhaag.com

A website for visitors and tourists, maintained by the The Hague Visitors and Convention Bureau. It contains information on events in the region as well as links to hotels, restaurants and cultural institutions. It does not offer information on public transport or parking and does not allow online reservations. However, clicking on 'The Hague for business visitors' brings up some very general transport information on how to get into the city by car or by train: here we also find links to the websites of the national railways company. The site has several technical flaws – most dramatically, the cultural agenda is empty.

Www.Residentie.net

This site has already been discussed at length in a previous section. It offers a wealth of information about events, sports, etc. The information is accessible not only to members of the Residentie.net community (you need to supply a login name and a password to become a member) but to any visitor. Unlike the passport access system, the webite does not allow for personalized information provision or alerts. Online reservation and ticket payment for theatres, museums or events is not possible. Furthermore, it fails to deliver online traffic information. As discussed, the site's major flaw is its failure to serve as the digital platform for all The Hague citizens. The site's content is maintained by an editorial board of five people: KPN, the former Dutch telecom monopolist, provides the technical maintenance.

In our frame of analysis, we state that vital local web content should be organized properly – that is, demand-oriented – in order to achieve the best results for the city. The Hague's content is presented in a very fragmented way. For visitors in particular it is difficult to find the information needed. The choice of separate 'portals' for municipal information (www.denhaag.nl), foreign companies (www.thehague.nl), residents (www.Residentie.net) and tourists (www.denhaag. com) is not convenient for the user. Furthermore, several of the sites have serious omissions or lack interactive possibilities. The weak content organization stems from a lack of vision on how The Hague should be presented on the web. It is also a result of weak cooperation, both public–public and public–private.

6 Governing Access: ICT and Internet Access Policies

Access to computers and the Internet is a central part of The Hague's strategy. The city realizes that in order to get 'Every Citizen Online', which is the central slogan of The Hague's e-strategy, it is necessary to ensure access for all. Another ambition is to improve Internet access among SMEs (90 per cent of the companies of The Hague belong in this category). This is not only needed to let all citizens benefit from the information revolution. Our 'digital flywheel hypothesis' suggests that higher levels of access also enlarge the local online market and will thus contribute to extension of online commercial activities as well as potentially forming a sound basis for the introduction of new services. Currently, 64 per cent of the city's population has access to the Internet and 75 per cent to a computer (see Table 9.5). This roughly equals the Dutch national average.

People who lag behind are the elderly, immigrants and people in the lower income bracket. The city has taken various initiatives to get more people online. Many of them can by now be characterized as 'traditional', for instance putting computer terminals in community centres and libraries, or offering computer and Internet courses at reduced fees. The city also benefits from national programmes to improve e-literacy. A notable example is the Digital Playground Programme, which supports the setting up of computer centres in deprived neighbourhoods, elderly homes, public places, etc. Computer availability in primary schools is still a problem. Our interviews suggest that perhaps a more important problem is the lack of skills of teachers to use them, as well as the lack of high quality educational content.

In The Hague, the library is one of the organizations involved in access policy. Halfway through the 1990s the library was very early to establish a community

Table 9.5　Access to the Internet in The Hague

	1999	2000	2001
Access to computer (home, work or school)			75%
Access to Internet	49%	57%	64%

Source: Den Haag (2001c).

computer centre in a deprived area of the city, the Schilderswijk, in partnership with telecom provider KPN. This was one of the first computer facilities of its kind in the country. Currently, the library has extended its activities. Outside office hours, the libraries open their computer rooms for schools and other organizations, and offer a number of computer courses. From 2001 a shift was observed in the type of courses demanded: from traditional 'how to use the Internet' courses and Microsoft applications there was increased demand for more sophisticated knowledge on how to create and maintain a website. Small companies and non-profit organizations particularly participate in these classes. As already discussed, there is no link to the Residentie.net initiative. This should be regarded as a missing link.

7　Governing Infrastructure

The city of The Hague is well endowed with ICT infrastructures like cable and copper. KPN offers relatively fast access through ADSL (asymmetric digital subscriber line, using conventional telephone line). Casema offers Internet access through the cable.

The market for high-speed Internet access in The Netherlands has a few distinctive features. First, there is little competition and second, there is hardly any 'real' broadband – through fibre optics – available for consumers. In the cable market there is lack of competition, as the cable network owner is also the main operator. Furthermore, cable offers high capacity for downloading but much less for uploading. This may become a problem in the near future when more and more people want to upload huge amounts of data, e.g. home video movies. ADSL reaches the limit of copper capacity, so hardly any future growth will be possible. In this field also, competition is restricted as KPN is both the owner of the copper infrastructure and the major operator. Although the company formally has to open its networks to competitors, in practice there are still many barriers to new entrants.

An opening in this deadlock can be fibre-to-the-home (FTH): it increases capacity almost infinitely and also easily allows for competition between operators in the infrastructure. The city already has a gridlock of fibre lines, connecting major offices (government and business) and business parks. However, there is no operator or consortium that invests in 'fibre-to-the-home'. For one thing,

Table 9.6   Speed of connection of different modes

	Download	**Upload**
ISDN	One way: 128 Kbps Both ways: 64 Kbps	One way: 128 Kbps Both ways: 64 Kbps
Powerline	1 Mbps–2 Mbps	1 Mbps–2 Mbps
ADSL	6–8 Mbps max	640 Kbps
Cable	27 Mbps	2.5 Mbps
Fiber optic	50 Mbps–20 Gbps	50 Mbps–20 Gbps

Source: BDRC Ltd, August 2001, appendix B.

these investments can be costly, notably when much digging is needed. Secondly, existing telecom operators are not very eager to 'cannibalize' their own markets by investing in fibre-to-the-home. Third, the expected demand for FTH is low, as the number of services that need high speed access is very restricted. This is referred to as the 'chicken and egg' problem: as long as there is no fast infrastructure, no broadband services will be developed, and vice versa.

Several Dutch cities are now trying to solve the problem. Eindhoven does so by subsidizing subscribers of broadband in the Kenniswijk, a test area in the city. It simultaneously promotes the development of broadband services, to make the infrastructure useful. Keeping the price low and promoting service development will increase the number of potential clients: this encourages companies to invest in fibre-to-the-home. A consortium is now planning to connect a number of homes. The City of Almere is developing alternative policies, involving housing corporations and communities of users.

For cities, it can be advantageous to have fibre-to-the-home. Demand for broadband will continue to rise and cities that are 'connected' are likely to reap most benefits. Recently, the city has put the issue high on the agenda. It has commissioned a study to research the feasibility of connecting every household to an fibre-optic connection. The idea is to create a public–private partnership in which the city, the major network owners (KPN and Casema) and some other actors together will invest in a city-wide optic fibre-optic infrastructure. Although the infrastructure would be a natural monopoly, access to the network should be open to any service provider, so that competition is encouraged on the service level. Service providers or other companies could lease capacity at market rates. In the current plans, the municipality would only be involved in the funding of the network, and not carry any operational responsibility of risk. It has been calculated that the costs of connecting every household would amount to €1,450 per household, €850 of which is for the construction of the 'passive' infrastructure. At the time of writing, a working group in which all the actors are involved is discussing the business model.

8   Conclusions

In our research we make a distinction between content, access and infrastructure as interlinked elements of urban ICT governance. We argue that organizing capacity is needed to optimize all three elements separately, as well as to connect them to one another. Organizing capacity includes, among other things, the presence of leadership, a broadly shared vision elaborated in strategies and high-quality networks in the city to get things done and to link resources and initiatives in a meaningful way. With regards to leadership and vision, we found that in The Hague ICT is high on the political agenda. At the highest levels, the ICT vision is supported by the mayor and aldermen and the city is investing substantially in realizing its ideals. The urban ICT vision guides investments and serves as lead for projects. However, what we miss is a view on the provision of electronic infrastructure, notably broadband. We suggest the development of initiatives for fibre-to-the-home. A comparative study on business models for fibre-to-the-home in some leading European cities could be a first step.

In the field of e-government, The Hague has high ambitions both to improve municipal service provision by bringing services online and to make the municipal processes transparent for the clients. Accordingly, it took some steps to organize itself internally and integrate the ICT systems: a strategic alliance with Microsoft was announced in this context. At the time of writing, Microsoft's role seems less strategic: apart from the 'normal' office software, it supplies the .net technology that is used to build primary process systems of some departments.

The ambitions of the Transparent City Hall project have not yet been realized: progress is slow and it remains to be seen how successfully it will eventually be implemented. Meanwhile, it has become clear that it is very hard to implement the envisioned information architecture. There seems to be a lack of support for reform in lower levels in the organization. Furthermore, e-government is a catalyst for new relations with stakeholders in the city. The 'marriage case' discussed above revealed a lot of uncertainty as to the role of the municipality, particularly the division of tasks between the city, the citizens and the private sector.

The most innovative project in The Hague is Residentie.net. The core idea behind Residentie.net is to promote the creation of a virtual city, among other things by making it easy for citizens and local organizations and firms to open up their online squares. We found that the Residentie.net project does not operate as expected. Visitor numbers are low and the formation of 'digital squares' falls short of expectation: participation in online debates is modest and the participation of private companies and local organizations is disappointing. Digital shopping malls have not come into being. Evidently, Residentie.net is now not functioning as *the* virtual city of The Hague. One of the key reasons is that the technology supplier in the project, KPN, has been given a role that does not fit its competencies, namely the exploitation of the Residentie.net. Another issue is the lack of cooperation with Digitale Hofstad, an existing local community of heavy Internet users. The lack of their support has been costly. Third, there is no link to the library courses. Fourth, there is little commitment in the town hall. In particular, the importance of online discussions is underestimated. Key partner KPN will probably withdraw at least partly from Residentie.net partly on financial grounds and partly because it has changed its strategy away from content. The question is, who fills the gap?

We conclude that The Hague's presence on the Internet is fragmented. There is no single and comprehensive portal where you can enter the city and find your way, whether you are a citizen, a tourist, a company or a commuter. At www.denhaag. nl, which would be the most logical point of entry, you only find information on municipal services and events. The English language version, www.thehague.nl, only offers information for inward investment.

Our analysis suggests that there should be an attractive, recognisable and comprehensive starting point, a 'portal' for anyone who wants to know anything about the city. Residentie.net should become an integral part of it, as 'My The Hague'. Major subsites would be the virtual city hall, the inward investment site, and so on. The same holds for the virtual legal capital. This should not become yet another portal, but part of the central portal www.denhaag.nl. This central point of entry could become a magnet for advertisers, as many hits can be expected. This homepage should be maintained by an independent organization with sufficient resources and knowledge to maintain the page technically, to keep it up to date – regional news, etc. – and, importantly, to create networks with all relevant organizations in the city.

The role of the municipality should be to provide its online content (services, information, news, etc.) as input to this organization. The website of the City of Hamburg can serve as an example. The portal should be managed appropriately. In the current climate, it is virtually impossible to make the exploitation of such a portal profitable: the city will need to invest heavily in it and also finance the unprofitable end of the project. Given the specific abilities that are needed to run such a complex portal, we suggest creating an independent organization. This public–private company should enrol editors able to organize and communicate current content in a comprehensive way.

Where electronic infrastructure is concerned, the current plans of the city to provide fibre-to-the-home are the most eye-catching and ambitious ICT plans of the city. If the city managed to create a network together with private actors and make it an open network in which there is competition on the service level, this would be quite an achievement. It is important to note that whatever business model is opted for, the costs to the taxpayers will be substantial. To justify this investment, the benefits should outweigh the costs. The benefits of having a city-wide fibre-to-the home network could be increased attractiveness of The Hague for citizens and companies, improved quality and variety of electronic services and more innovation. However, one of the key arguments for the city to invest in the network is that it would bring substantial benefits in the (semi)public domain, for instance in education, health services, security, culture and government services. Here, one critical point should be made: *the extent to which these benefits will indeed be realized depends to a crucial extent on the ability of the urban actors in these sectors (i.e. hospitals, police, schools, local government) to create useful and meaningful broadband content.* We have seen in this study that actors in The Hague have problems in creating even 'narrow-band' content (witness the case of the 'marriage portal'). It requires not technology, but rather organizing capacity, i.e. the capacity to cooperate in networks, to share knowledge and information in order to create new services and solutions. This will be the key challenge for the city. In our view, high investments in electronic infrastructure can only be justified if the city improves its performance on the content side.

References

BDRC (2001), *The Development of Broadband Access Platforms in Europe: Technologies, Services, Markets*, 1 August.

Den Haag (2001a), *Het Glazen Stadhuis: De Haagse aanpak voor burger en bestuur, Projectplan 2001*, Superpilot BZK.

Den Haag (2001b), *Den Haag, Stad in de Informatiesamenleving, Informatiewerkplaats*, RIS70390a.

Den Haag (2001c), *Stadsenquete Den Haag*, City of The Hague.

Shapiro, H. and Varian, H.R. (1998), *Information Rules: A Strategic Guide to the Network Economy*, Boston, MA: Harvard Business School Press.

Interview Partners

C. Beekhuizen, Department of Human Resource Development, Organization and Information, City of The Hague.

R. Buitenman, De Digitale Hofstad.

F. Dane, Project Manager Residentie.net, Department of Human Resource Development, Organization and Information, City of The Hague.

A. van Duuren, Department of Public Library Services, City of The Hague.

F. van den Eerenbeemt, Director, De Informatiewerkplaats.

R. Koene, Project Manager Residentie.net, Casema.

J. Overdevest, Project Manager Residentie.net, KPN.

M. van Rossum, Management Consultant, CEO Infocities office.

M.J. Snels, Department of Human Resource Development, Organization and Information, City of The Hague.

W. Stolte, Alderman for Information Policy, City of The Hague.

F.A.F. Toet, Department of Human Resource Development, Organization and Information, City of The Hague.

W. Vroegindewey, Sales Consultant, Microsoft.

Chapter 10

The Case of Venice[1]

1 Introduction

In this chapter, we will describe and analyse Venice's e-governance strategies. We will start in section 2 with a general context description of the enormous challenges that the city of Venice faces. This puts the city's e-governance efforts into perspective. In section 3 we summarize the city's ICT strategy. In sections 4, 5 and 6 we analyse the issues of content, access and infrastructure respectively. Section 7 concludes.

2 Context

Territory and Demography

Venice is the capital city and administrative centre of the Veneto Region, in the heart of the Italian northeast, a macro-area that represents a strong economic subsystem with peculiar characteristics. The Municipality of Venice extends for 189.4 km^2, partly in islands located in the middle of the lagoon, with the historical centre occupying a major cluster of islands totalling 7.6 km^2. The rest of the municipal territory comprises a section facing the sea ('littoral', two main islands plus an inland neighbourhood that is now seceded) stretching 48 km^2 and a main inland section of 134 km^2, which includes a middle-sized city in its own right, Mestre (150,000 inhabitants) and Marghera, a smaller town but a major industrial and port area. The municipality of Venice with its main administrative articulations is described in Figure 10.1 Venice is a well-known international attraction, possibly the most famous tourist city in the world. Yet it is hardly known that its historical centre (henceforth Venice historical centre or Venice HC) in the heart of the lagoon is a 'problem area', whereas its unattractive inland settlement is well integrated into a booming regional economy. With young households pushed out of the centre by inaccessible housing prices and lack of specialized jobs, the population in Venice historical centre went from 170,000 to 70,000 in around half a century.

Figure 10.2 describes the demographic evolution in the last half a century. The population path follows a typical pattern found in many European metropolitan

---

[1]   This chapter is written by Dr Antonio Paolo Russo.

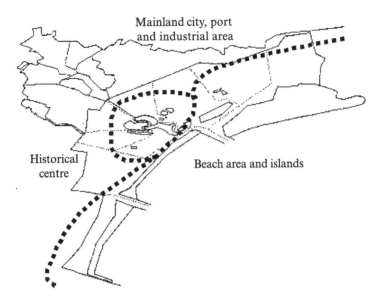

Figure 10.1 Municipality of Venice with main subdivisions

areas (cf. Klaassen et al., 1989). The historical centre has been in decline since the end of World War II at an average yearly rate of 2 per cent, moving towards 1 per cent in the last decade. The inland areas replicate the general pattern, stabilizing at around 170,000 inhabitants in the last two decades: the littoral kept its size almost constant at around 50,000, dropping to 35,000 after the secession of Cavallino neighbourhood in 1999.

Environmental problems undermine the very durability of the city, chasing away scared investors, inhabitants and economic operators who bear the high costs of environmental degradation. The 'high water' (flooding) – increasing in frequency and impact due to the on-going erosion of the lagoon bottom – may be an unexpected thrill for curious tourists, but it is a real drama for households and economic activities. Tourism cannot be held solely responsible for the deterioration of the delicate lagoon environment: rather, it is the result of global phenomena (the increase in the sea level and air pollution) and inappropriate decisions in the past. Yet it certainly contributes to worsening existing trends. The limited number of entry points to the old city causes structural traffic on the only road link to the mainland, high levels of air and water pollution, plus heavy stress along the main pedestrian connections between such points and the main tourist area of St Mark.

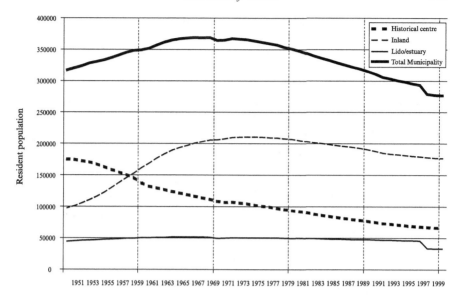

Figure 10.2 Population structure in different sections of Venice Municipality, 1951–1999

Economy and Administration

Whereas the historical centre and the smaller islands surrounding it concentrate the beauty and wealth inherited from Venice's millenary status of main centre of commerce, artistic production and military power of the Mediterranean, today the economic heart of the region is found in its 'back garden'. The metropolitan area nicknamed 'Pa-Tre-Ve' includes the city of Venice and extends to the neighbouring province cities of Padua and Treviso. The former is an art city and main service centre, the seat of one of the most prestigious Italian universities. The latter is a smaller historical city, mostly famous for being at the heart of an industrial district (Benetton's and Stefanel's headquarters and main factories are located here) with a very distinct image in international markets. However, no formal administrative authority corresponds to the metropolitan region. The Region Veneto has competence over most matters regarding infrastructure and investments, while the smaller territorial unit of the Province of Venice has competences over employment and tourist promotion. To this complexity, the role of the state in innovation and university education must be added and that of different national authorities, such as the port authority, over large portions of the city assets. Finally, many private investing groups and international agencies are active in the region. Dente (2001) counts 18 agencies consistently involved in project developments.

The economy of the urban region is gradually shifting away from the port-driven industry. Since the 1930s, when the city could count on an industrial area of prime importance and location, many things have changed. The restructuring of the Italian heavy industry and the huge environmental problems caused by petrochemical productions have caused a downward drift in Porto Marghera's competitiveness. Fortunately, rather than provoking a real crisis in employment, this decline has gone hand-in-hand with a spatial dispersion of economic activity and a structural break in production, from heavy, polluting and state-supported industry to a highly competitive district economy driven by SME. For this reason, the unemployment rate in the area continues to be one of the lowest in Italy, reaching full employment in some municipalities and averaging 5.1 per cent at the provincial level. The abundance of employment opportunities attracts many immigrant workers to the area, especially from North African and Eastern European countries, with minor socio-cultural impacts compared with other Italian regions characterized by strong in-migration. The provinces neighbouring Venice are part of the so-called 'economic miracle' of the third Italy – although with distinctive characteristics and with emerging problems that could cause a loss of competitiveness in the long term – the most important of which is the inadequate quality of the regional infrastructure and human resources. Despite such worrisome perspectives, the 'northeast' economy is still regarded as a success story.

Challenges

Despite its indisputable weight in the local economy (without considering indirect multipliers, tourism accounts for 30 per cent of jobs and 24 per cent of economic units in the historical centre), tourism is blamed as the main culprit of the city's economic and political decline. At the same time, the city seems unable to survive without it or to embrace other opportunities.

The historical core of Venice is facing problems in finding its role. Excluded from the 'large network' of international trade that made the 'Serenissima' republic great, and now in constant peril of being disconnected from the booming backyard economy which is anathema to its ancient splendour, Venice experienced a meagre growth in services economy around 1 per cent in the decade 1981–1991, despite growth increases of 20–40 per cent in the surrounding cities (Rullani and Micelli, 2000). This trend persists and today, one of the problems that the local government has to solve – apart from the day-to-day problems caused by tourism, environment and physical degradation – is how to 'reinvent' the city's economic vocation, providing a solid alternative to the arguably volatile and environmentally-destructive tourism economy.

The recent change in the political environment, with the direct election of a mayor and city government (who are granted in this way more responsibility and authority) has fuelled a wide-ranging debate on strategies to reverse the existing trend and initiate a 'virtuous cycle' of development. The assumption is that the structural barriers that have kept old Venice disconnected from the economic boom of its hinterland can be removed with a modern and comprehensive infrastructure policy. Once this is achieved, the city can count on plenty of complementarities with its region. Moreover, in Venice there are under-utilized locations and other 'non-material' resources that can be put to good use. The mobilization of this potential would allow Venice to re-establish a role in the network of the globalized economy, with a favourable position at the heart of an area – the Alpine-Adriatic – that has become a critical node in the new European commerce flows. Initiatives in this area do exist: the regeneration of waterfront areas, like the famous Arsenal and the Giudecca Island, for maritime-connected functions; the launch of research and knowledge-dissemination centres like the Science Park VEGA and the Venice International University; a new impetus for modern architecture and planning; and a renewed effort to market existing resources to international investors. These projects offer development opportunities and contribute to the refurbishment of the obsolete image of Venice as a production and investment location.

The valorization of the enormous stock of cultural assets – and the very cultural atmosphere of Venice – is a key challenge in this respect. Cultural tourism has the potential to generate the value that is needed both to preserve the heritage and to foster a new cycle of development, based on culture-intensive and 'intangible' knowledge. However, this is not the case. To say the least, the potential offered by culture is not properly exploited by mass tourism; this model affects the very integrity and durability of such heritage. Once the worst effects of tourism mismanagement have been eliminated through careful regional and spatial planning, tourism could indeed become a lever for the economic regeneration of the city, but to do so, it needs thorough restructuring.

The vision of Venice as the 'meeting place' of an economically thriving region, infrastructure node and education centre and capital of culture and artistic production, patronized by well-educated and informed visitors, is gaining ground among the circles of scholars and forward-thinking politicians. However, the biggest challenge today is to involve in this process a larger part of the population, who are increasingly disaffected with a city that is seen as beautiful but uncomfortable and expensive for everyday life. Venice is a complex city to understand, semantically. It is necessary to 'guide' the population about the resources existing here. Therefore, regeneration strategies also have to address such issues as inclusion (both in terms of participation and integration between 'use-systems' of the city's resources), education and 'internal marketing'.

3   Venice's ICT Vision and Strategy

In this section, we start our analysis of Venice's e-governance performance. We will first define exactly the problem that the city is facing. Finally, we will analyse how the city can meet these challenges by developing an e-governance strategy, and the role of the stakeholders involved in this process.

Problem Statement

Today, Venice has a number of problems with tourism. First, the pressure from visitors' flows is excessive. Excursionists, that are the least manageable and the most 'costly', invade the city on a daily basis in peak days, provoking a substantial paralysis of other vital urban functions. Research (Russo, 2002; van der Borg, 1991; Costa and Manente, 1995) has pointed out that the presence of a large share of excursionists is partly due to the limited amount of accommodation in the historical centre and partly to the structure of the tourist market in the destination region. As a consequence of this, visitors are unwilling to spend much in the city and especially to pay for culture – they just wander around the city. This derives both from the short time available for the visit and the overcrowding of the main, central attractions, but also from the lack of information on existing peripheral assets (Caserta and Russo, 2002). Finally, the tourist economy and the rest of the economy seem to travel on different rails (Rullani and Micelli, 2001). Much of the tourist revenue escapes the local economic circuit as it remains in the hands of outsiders. Moreover, there is a high rate of tax evasion on tourist revenues. Thus the tourist economy 'chases away' the commercial and employment structure of the city that is primarily serving the resident community. At the same time it is subject to processes of quality downturn in the face of a demand that is merely experiential and characterized by limited information.

The issue is, then, how to enter a new stage of tourism development, one that enriches the city and creates opportunities for synergies with other strategic sectors of the city – culture, but also knowledge-intensive industries, residence, education, etc. This also means changing the way in which the city is visited, through careful marketing, segmentation and pricing strategies, awareness-creation and the enhancement of the 'systemic', value-generating nature of the cultural enterprises.

Public actors are pre-eminent in the process of adapting ICT to manage tourism better. The relevant region for Venetian tourism – the one that is actively involved in the generation of tourist flows towards the city – spans a whole portion of the northeastern Italian territory (including a region, seven provinces and countless municipal administrations), at least three neighbouring regions and adjacent areas of Austria, Slovenia and Croatia.

In general, tourism is an exemplary illustration of the necessity to bring forward innovative governance models to guarantee sustainability in the development process. If there is no cohesion and agreement on the goals, private operators as well as visitors will follow strategies that are basically detrimental to the destination's long-term competitiveness. Therefore, the complexity of objectives attached to tourism development in a region has to be fully understood. There are at least three different communities involved in the production, delivery and consumption of tourist commodities: the private sector and its organizations; the citizens and democratic institutions of representation; and the visitors themselves. A 'horizontal' typology can then be proposed according to their degree of 'localism', that can be the very local (e.g. the level of municipal administrations), the meso-local (regional, provincial or national) and the no-place or global level that increasingly characterizes the international economic activities like the tourism industry. In this way, a nine-entry matrix can be reconstructed that covers the whole range of stakeholders involved in the tourism system (Figure 10.3).

		Actors		
		Citizens and institutions	Private sector	Consumers
Place reference	Local	Local population City council	Local business community Chamber of Commerce Hotel association, etc.	Local cultural consumers
	Regional	Population of neighbouring cities and provinces Provincial and regional governments	Regional tourist board Transport companies	Daily shoppers Schools Regional cultural consumers
	Global	World population, ethnic and cultural groups with roots in local community	Travel and tourism industry corporations	National and international visitors

Figure 10.3 Stakeholders in a local tourist system

Source: Based on Russo (2002).

This typology reflects a diversity in goals; for instance, between the community of Venice that is interested in keeping tourism under capacity thresholds, and a neighbouring municipality that free-rides on its proximity to Venice and is clearly (and blindly) interested in a model of mass exploitation of the central resources. In the face of this, the response in terms of governmental capacity is not the most adequate one and, moreover, it is very difficult for tourism policy – mainly carried out at the regional or provincial level – to 'hook up' with local development initiatives. In terms of ICT adoption, with particular reference to tourism management, one should note that the agencies in charge of developing the content (APTs: Agencies for Promoting Tourism, local cultural-tourist industry), access (region, state), and infrastructure (municipality, state, private operators) do not always share the same goals or have consistent strategies in term of destination management.

Towards an E-Vision and an E-Strategy for Sustainable Tourism

A key problem in building a sustainable tourism is the lack of cultural and organizational empathy among: a) tourists; b) tour operators; c) local tourism/ cultural organizations; d) local hospitality providers (Go, Lee and Russo, 2002; van der Borg, Russo, Minghetti, 1998). The ones that deserve special attention as the most challenging are:

- **a ↔ c**: how can tourists be brought to greater cultural awareness and appreciation?
- **d ↔ a**: how local hospitality providers (in particular SMEs) can get a better marketing empathy for their clients?

Beyond diverging economic interests and differences in roles, it is believed that mutual understanding can be increased utilizing the proper educational and relational tools, which may exploit the possibilities offered by information and communication technologies (ICTs). The multi-system nature of ICT appears promising as a vehicle to stimulate positive side-effects to tourism development and to improve the coordination of the involved actors and institutions. It is crucial to this aim that the ICT apparatus is diffused to the tourist sector as a whole, connecting cultural producers and linking the local network to the tourist industry.

ICT applications may be developed at various levels. In the framework of Buhalis (1997), ICT re-engineers intra-organization, inter-organizations and customer relations functions bringing the three dimensions closer to one another with enormous impacts on costs, productivity and quality.[2] Thus, ICTs facilitate both front-and back-office operations. Restructured back-office operations, where information can be exchanged *along* and *across* chains, may lead to a more cohesive and integrated industry. Restructured front-office operations enable a

---

[2]   By allowing a direct contact via web, information kiosks or emerging intermediation platforms like interactive digital TV and mobile technology (Buhalis and Licata, 2002) between networks of producers and customers, ICTs allow a smart packaging and marketing of the destination. In intra-organization operations, ICT booking and information devices learn the profiles of visitors and their preferences, and this is used for a high-level segmentation of the market; the system suggests personalized products and itineraries, giving the opportunity for advance bookings, special offers and interactive information retrieval. Finally, back-office operations are made smarter as ICTs allow joint marketing initiatives and coordination between firms, through systems that support integration at different levels of the chain.

higher discriminatory power both to the demand and the supply side. Firms are pushed to compete on uniqueness and *content* rather than cutting down quality to ensure higher profit margins. Unintermediated marketing relationships between producers and visitors erode the 'information' and 'location' rents on which most of the sub-economy of the heritage city thrives; small, peripheral operators can compete with larger, better located incumbents if they can offer more convenient products and a smarter packaging. In the end, the decisional scheme of heritage tourists is also restructured: secondary products can be chosen after, and as a function of, the chosen itinerary. Therefore, by packaging itineraries and cultural products, destinations can modify the location behaviour of the hospitality industry, leading to a more balanced tourist region.

It should not be forgotten how the development of a modern relational infrastructure is not only germane to a better management of tourism, but also has spin-offs for other sectors complementary to tourism and culture, such as the development of a strong knowledge-economy basis or the attraction of headquarter and administrative functions. ICT may foster the 'network' structure of a city that has so far proceeded in a strongly compartmentalized development process. By investing in soft accessibility and processing power, Venice can overcome much of its objective limitations, like the physical isolation, its reducing social base and its fragility, and instead boost its intangible points of strength, like its worldwide fame, its multicultural character and its status, that give it added value and make it a real 'small-size' global city. The idea is to promote Venice as a district for technological, cultural and societal innovation, hosting a number of eminent international institutions.

A number of recent projects have started a process of rationalization and restructuring of tourism. The application of ICT is making the difference in this new approach, enabling a smarter, wiser and ultimately more sustainable approach to visitor management. This is based on the same technological platform that also allows the cultural-tourist cluster to evolve into a more flexible and effective organization. This double layer of restructuring is proceeding at different speeds, with front-office applications opening the path and restructuring of back-office operations expected to follow. A first step in this direction was taken by the municipality, which has formed a public–private partnership for the electronic ticketing of four museums, the Cathedral and the Tower in St Mark. Even if limited in scope, this project opened a path which, if successful, can easily be extended to other levels of flow regulation, such as transport, terminals and the hotel and restaurant sector.

These first steps are currently being articulated in a real e-strategy for Venice and its tourism. This is emerging as an array of initiatives that connect in various ways, rather than as the result of a clear-cut programme. In fact, no real e-government project was carried out in Venice until recently. In Italy, after an initial moment when the potential (and the limits) of new technologies had to be recognized and accepted by public administrations, some proactive municipal administrations started to explore the practical possibility of using ICT in order to improve their efficiency and the quality of their services to citizens. The two most advanced experiments of 'civic networks' were set up in the early 1990s in Milan and Bologna. Venice hooked up with lower scale-projects at a later stage. The Italian telecommunication company – formerly operating in a monopoly regime – endorsed the idea of Venice new 'pole of telecom', investing in a top-class research centre right in the middle of the historical city and starting a wide-ranging programme for cabling the city. Moreover, they funded many important projects of social relevance. The main idea behind these efforts was to create online communities to support specific social and economic issues. An example of a pilot project regarding the organization of 'web communities' through a city portal is the institution of a website to assist and support the carers of Alzheimer sufferers.[3] Another is the creation of a network of professional artists and art managers, in order to make it possible to overcome the existing barriers and operate successfully in the city.

The initial 'bottom-up' impetus has partly been lost, but the city is now keen on taking the lead in a revived and comprehensive regeneration strategy, of which the e-strategy is a building block. An agenda of initiatives is being formulated where *content*, and no longer infrastructure, is at centre stage, and ICT is the enabling tool. One example is the 'Venice District for Innovation'. This initiative will facilitate the settlement and development of innovative business activities: through it, the City Administration intends to attract new investors and entrepreneurs. Other initiatives, to be mentioned later, specifically target tourism and culture.

The 'brain centres' of this new stage in the regeneration policy for the city are in the City Council (the Mayor's Cabinet and the Office for International Relations and Special Projects): the e-initiatives that accompany it depend on a number of technical 'arms' like the municipal IT department and Venis as well as a wide range of private and public partners variously involved in the field of urban management. Some of these are described here.

Consorzio Venezia Richerche (CVR) is a non-profit venture putting together the local universities, City Hall and some of its public companies and large firms of both local and national relevance. The concept evolved into a science park, now located at the gateway of the city in a former industrial site, which is mainly active in knowledge transfer, education and start-ups. The activity of CVR covers four areas: environment, culture, ICT and materials (plastic and polymers with the petrochemical industry). CVR launched several relevant projects in the e-strategy of the city such as the Promemoria project, which was about the cataloguing of works of art, and the Calypso project on smart cards for citizens and tourists, described later.

ASM is a private company specializing in parking and mobility management on behalf of the City of Venice. They are also the main players behind a web-

---

[3]   This condition is relevant in an ageing city like Venice, where most health services are hard to access due to the structure of the city. The original association committee evolved into a website, originally supported by the city but now self-sustaining.

portal, Urbis Limen, providing advanced bookings for parking places in the city.

Azienda di Promozione Turistica (APTs) provide tourism marketing and management. APTs are agencies that cover more municipal territories within a province – Venice's APT includes some coastal resorts and the rural areas in Venice's 'backyard'. APTs have recently undergone a change in regime, passing from the regional to the provincial authority. The territorial reorganization of APTs was expected to bring a change in management and in the stakeholders' involvement, possibly creating a unique laboratory for networking. APTs will become private companies, controlled by the Province (51 per cent), city (14 per cent) and for the rest by other partners (trade unions, transport firms, union of commerce). The cultural sector will be represented in the system via the municipal department that runs the museums. The main municipalities in the province will also be present in the board as well as one Promovenezia, the catering sector, AVA (hotels), Assindustria (big hotels).

University of Venice – Ca' Foscari. The principal higher education institution of the city is mainly oriented towards social sciences and humanities studies. It includes four faculties offering a total number of 13 first-degree Courses plus 13 three-year diplomas (or 'short degrees'). The main bulk of educational activities is organized around two main poles: the S. Marta-Zattere complex, a former maritime-industrial site in the southwestern part of Venice, and the new S. Giobbe complex in the north side of the city, where the Faculty of Economics is almost entirely hosted. Some lecture halls are decentralized around town. The Information Science faculty is almost entirely located in the inland industrial area of Mestre, where it is intended to be a research stimulus for the advanced technological applications at the Science Park VEGA.

IUAV is an autonomous architecture university, one of the oldest and most prestigious in Europe. Similarly to Ca' Foscari, IUAV has been forced to spread its activities over a number of historical buildings. Recently, unused industrial sites have been taken over by IUAV for the more space-consuming activities. IUAV was born as an athenaeum of formation and research in the field of architecture; over the last 30 years a line of research and education in urban planning has also been developed. IUAV's three faculties of (Architecture, Design and Spatial Planning) offer five first-degree three-year courses and 11 specialized degree two year courses.

The *Venice International University* (VIU) is the most recent addition to Venice's higher education institutes. Born as a partnership between the two Venetian athenaeums and four foreign universities, an important banking foundation and the Province of Venice, it manages higher education and a number of research centres through undergraduate, graduate and continuing education and scientific research. VIU's educational activities are organized as a joint undergraduate programme in cooperation with the partner universities. Currently, two master courses (MAs) are offered, one in Economics and Finance and one in Cultural Mediation towards Investment and Integration. A number of research centres are based at the VIU. Among these, two can be singled out for a potentially fruitful contribution to local issues. *TeDIS* (Centre for Studies on Technologies in Distributed Intelligence Systems) is concerned with the impact of ICT on Italian industrial districts as a departure point for general considerations about the evolution of new technologies in an economic scenario, where SMEs networks play a dominant role. The siting of TeDIS and its experts in the northeastern economic community ensures the practical relevance of its high-profile research output. These actors are involved to various degrees in the development of the strategy of the city for e-governance. However, this strategy faces some problems regarding the coordination between local and national initiatives. The e-government projects that need to tap into the national budget are evaluated according to a set of parameters and need to be consistent with a regional plan. Until the time of writing, only four regions in Italy had presented regional plans, Veneto included. The goal of the government is to reduce chaotic and expensive competition in the infrastructure projects, but as a result many existing initiatives are downplayed and local specificities tend to be overlooked. These activities are described in the next section, where we will evaluate the potential they offer to a more sustainable Venice and their level of achievement, analysing their weaknesses and potential points of improvement.

4   Governing Content

We now focus more closely on the city's e-strategy. Tourism-related initiatives are centre stage. However, their added value would be, in a sense, their capacity to be valuable for a wider range of customer groups than visitors only. For this reason, it is important to analyse the connections and synergies that exist between different applications.

Web Strategies (Front Office)

General citizen services
The city is busy developing a city portal to offer information and services to citizens, following the general directives of the Ministry for Innovation. These identify those services of the state and local authorities that can be managed through these portals, such as online payments of taxes and fees. In the short term other facilities will also be available, for instance, access to the catalogues of the municipal libraries and municipal museums. The latter have international relevance, so the 'user group' is particularly large and interesting. In the medium term, the city website will develop as a territorial portal, a one-stop window for municipal services and other firms operating in the region. This portal will give access to services like patents for starters, new shops, building permits, health services, and fire brigades. A multiplicity of institutions cooperate out of a regulatory agreement with established procedures. They are also experimenting

with the digital signature as a 'registration authority', making it possible to deliver services on the web, give documents of certification and authorization, affecting heavily the efficiency of the offices and it is not always accepted by the bureaucracy.

A further step regards the project of decentralizing the municipal services, with the institutions of 'District Municipalities' offering front-office services and connected in the back office with the main City Hall administration. The experimentation started slowly because of staff problems, but is now progressing. The Municipality of Marghera, a working-class area located close to the industrial port, is among those that are likely to benefit most from these services. This community, affected by social and environmental problems, can really be seen as a pilot of a new model of involving and servicing the residents. Marghera has human capital capable of initiating a new generation of small and medium-sized enterprises in the service sector and access to e-government services can help this process. Education can also benefit from having access to the cable infrastructure available for the public services.

Tourism-related projects

With special regard to tourism management, City Hall has been the initiator of an interesting project of front-office facilities for tourism which is completely consistent with the idea that tourism has to be managed on a wide territorial scale, establishing partnerships with all the players involved.

ALATA (High-Adriatic Partnership for a Sustainable Tourism) was set up as a system for the management of visitor flow to the northeastern region of Italy on the occasion of the religious celebrations for the Holy Year 2000. It was feared that an excessive number of pilgrims travelling independently to Rome would cause intolerable congestion in the tourist destination. Venice was expecting an estimated additional flow of 3 million visitors motivated by the event, hence the opportunity to divert this flow to peripheral but well-equipped destinations with some cultural or religious attractiveness. Whenever the central destination reaches saturation, or other peripheral attractions have something to offer, this information becomes public and the incoming visitors can modify their itinerary accordingly, finding all the information that they need in one 'mall'. The rationale is that the entire world is interested in Venice, but not in its 'periphery'. By delivering the 'right' information on culture, events and facilities, a better spread of the visitor flow over a large territory could be achieved, relieving the pressure on Venice and instead creating the momentum for investment in culture and hospitality in localities that normally are just 'buffer zones' for the Venetian market.

In synthesis, the project's aim was the realization of a telecom infrastructure and software for the collection, management, certification and redistribution of information on visitor flows in the High-Adriatic territory, as an input to just-in-time provision of facilities for welcoming, assisting and accommodating pilgrims and tourists and facing possible emergencies. A service card was distributed to grant access to dedicated facilities at 200 ALATA points with an operator, located in places such as APTs and terminals, where you could obtain information, itineraries, etc. The back-office system consisted of eight control centres where data were input and a central operational office. The system connected the existing transport, hotel and catering structure realizing the 'links' and providing dedicated facilities. The database was in part 'institutional' (a complete list of prices, addresses, information on sites, etc.), and in part 'commercial' with a list of hotels and attractions providing online booking.

The project successfully achieved the multiple objectives of coordination at the spatial level and promotion of selected facilities and attractions, enlarging the spatial scale of Venice as a destination area and redistributing accordingly the management and investments costs of a tourist system focused on Venice, but involving a larger territory. The practical possibility of attracting consistent visitor flows to these 'peripheral' areas was one reason for administrations who had previously played a passive role in this system to become engaged from the point of view of the organization and management of the visits.

Even though the system had to operate in an emergency situation during the year of the event (which, however, attracted a lower number of tourists than expected), the ultimate aim of the ALATA partnership was to utilize this system in the 'normal state'. However, two years after the Holy Year, the momentum to keep actors with diverging interests together had vanished. The visitors flow went back to its 'structural' trend and is interested in little else but Venice. The motive to coordinate the system's core and periphery is less strong and, although it would be necessary to continue this experience in order to achieve a more sustainable tourist system, the partnership has virtually disbanded. The front-office activities of ALATA no longer receive funding and back-office operations are hardly seen as necessary. The reasons for this unfortunate outcome can be traced to the political problems of coordination between municipal administrations with different political colours, which had financed the system through their ordinary budgets. At that point, everybody would rather promote their individual place rather than the whole system. Moreover, in 1997 the Internet was not so diffused, whereas currently it is easier for small private operators to set up an information and booking system which is more competitive than ALATA. At the time of writing the actors remaining in the ALATA partnership are trying to reposition this service. ALATA provides the operation and booking system for the Venice card and other local services, but obviously its potential extends to a larger territory. The challenge is to make it become commercially self-sufficient and this takes place through a market-mediated relation with the operators in the system. However, in this new model ALATA risks losing its main advantage – the public-sector services market – which means selling, for instance, museum services online. This product is one that can really 'bend' the tourist market.

A second project that focuses more closely on the 'cluster' characteristics of the cultural tourist system is *Venice Card*. This project had a long gestation. Originally it developed as a project called CALYPSO, financed by the European Community (DG XIII). It foresaw the realization of a smart card, providing a number of services to citizens in a coordinated and user-friendly way, integrating payments and banking services, urban transport, student services, any kind of bookings and information to city users. The card was supposed to work as a 'pass' (utilizing 'contactless' technology) to access a number of facilities and functions. CALYPSO experimented in four European cities (including Paris, Konstanz and Lisbon). In Venice, partners in the project are the main public and private operators in the field of transport, banking, municipal services, the universities, the museums and a consortium that manages the religious heritage. The development of this project foresees the issue of two types of card, one for local users and one for tourists, which can be 'charged' with services and electronic money when visitors book their visit and receive the card.

The number of cards issued is to be equal to the tolerance threshold, to be periodically determined. In this way, motivated cultural tourists receive a better deal because they can more easily discover what is on offer and arrange their itineraries according to benefit from free parking, access to limited-number events, reduced time in queues, reductions in transport fares, etc. Meanwhile, the city is better off because it attracts relatively high-spending and organized tourists. Incentives for advance bookers represent a feasible and socially acceptable formula for having the visit to a heritage city paid for (Di Monte and Scaramuzzi, 1997), which, combined with telecom facilities, yields an intelligent way to selectively market the city and spatially/seasonally smooth the peaks. Overcoming the barriers to entry would make it possible to draw on the position rent on which such a large part of the Venetian tourist sub-economy lives, thus increasing the quality of the tourist experience.

The tourist card, called the Venice Card, was eventually introduced in 2001. Venice Card focuses on the promotion of specific types of cultural consumption and their integration with other aspects of the service economy. The Card is reserved by tourists when they send in the booking for their accommodation and physically collected upon arrival at the city's terminals. However, the first version of the card was not a 'smart card', but a normal ticket that works as a marketing tool but hardly achieves any logistic/management objective and is insufficiently flexible for different kinds of tourists to build their own itineraries in an interactive environment. This is due both to the difficulty of having all the relevant actors participating in the system right from the start and to associated delay in laying down the necessary infrastructure. The Venice Card managing company therefore opted for a 'low-profile' introduction, hoping to build the consensus for the 'smart' version of the Card in a learning-by-doing approach. However, the process of integration is not without problems, as it will be seen later.

These projects tackle some of the dimensions of tourism most likely to be subject to unsustainable development, namely the spatial and the industrial level. Basically, ALATA spreads tourism in a more efficient and rational way on the territory, realizing a soft demand regulation and associating it with high quality services. The Venice Card favours an 'integral' approach to the management of cultural tourism, integrating the infrastructure for visitor service, lowering the 'information barriers' which foster quality decline and granting access to electronic mall to non-standardized cultural production and events, side-stepping the bottleneck represented by an intermediary sector which is not prone to invest in 'novelty'.

One last field of application of ICT which is supposed to enhance the city's economic performance is cultural management. Venice's *e-culture strategy* started some years ago with the 'Promemoria' project, a business model for local museums. Municipal museums had serious problems in setting up a technological platform for marketing and selling their content more effectively. Moreover, the publicly-managed institutions lacked a real vision in this sense and a practical sense of the opportunities from the application of ICT in a cultural sphere. Without a specific strategy in cultural communication, the segment of visitors who are more interesting for the city – those who are really interested and willing to pay – cannot be reached and the tourist market is hardly differentiated. The challenge instead is to get people interested in a specific cultural product together and to have a communication strategy directed at them, thus developing a users' community. Proactive institutions still perceive technology as a threat and do not invest in it. After a while, it was clear to the managers of Promemoria that an effort to familiarize potential users of cultural management with ICT had to come before any content could be developed. This had a slow impact and could not count on any institutional backing.

At the time of writing, the contents can be developed and the discussion has started on how to put e-culture to the service of the city, contributing to a sustainable development. ICTs promise to increase the performance of museums and collections through smarter internal operations, commercial services and advanced communication strategies. For instance, the Municipal Network is thinking of using ICT to sell online catalogues of works of art and to appeal to different market segments, for instance children. This group is difficult to involve, but represents a potentially interesting and large share of users. ICT can greatly aid the interpretation and transmission of the cultural content to this group. ICT also provides seamless access to resources that one can physically visit in the city, as well as live experiences regarding the history and environment of Venice. In this way, the quality of their tourist experience improves enormously, as they are able to establish connections and organize their real visit around various themes and suggestions, escaping in this way the 'tourist bandwagon' cliché. The 'virtual heritage' has the potential to affect in the desired way the logistic structure of the visits and the behaviour of visitors – for example, willingness to pay for cultural resources and the return patterns. This project may also bring forward a new attitude of visitors towards the cultural heritage of Venice, increasing its attractiveness and comprehensibility, and therefore its capacity to generate value, while at the same time making tourists more curious and less predictable in the organizations of their cultural itineraries.

Internal ICT Systems (Back Office)

A necessary step in the city's move towards a smarter management of tourism and other urban functions regards the rationalization of services for different categories of city users including not only residents and visitors but also students, daily commuters and other important population groups like the elderly and foreigners. This requires a reorganization of back-office operations at city level. The municipal administration is going through this rather painfully, having had to adapt to a fragmented organization framework.

The City Hall has two distinct directions engaged in ICT: the Municipal Network (Communication Office), which has formal competence over the contents of front office operations and the Information System Department, mainly involved with back-office development and coordination. The cooperation within these two departments and with all the other departments concerned with the municipal operations (e.g. culture) is not without problems. The municipal offices have started experiencing new standard procedures to deal with files. A peculiar service in Venice is the management of the municipally-built stock. The data on tenants and renovation projects form a database that is 'retraceable' by users on the web. One of the most contested issues is the identification of the steps of the files from one workstation to the other, which requires an understanding of acceptance of the file. The file is managed through the construction of a virtual file where all documents and annexes are archived. This allows users to know which documents are missing. Such pilot experiences are bringing forward a cooperative climate in the City Hall, as unskilled workers see their utility and realize that they do not have to be afraid of user-friendly and time-saving new technologies.

Integration

Integration between different ICT projects has to be achieved at two different levels: the integration of internal platforms and procedures for back-office operations, and the integration of parallel front-office projects. Both these layers presented a number of difficulties that caused a serious delay in the process of ICT adoption.

Venis Ltd is a semi-private company (its budget is a part of the budget of the city of Venice) in charge of integrating the various information systems utilized by the Municipal Administration and delivering ERS services. Until recently they were the only provider, with total responsibility for the city services. Now they are fully responsible only for system management and integration and compete with other providers for new systems and projects. In this transition, Venis had to change its mission. Now they undertake the not-for-profit task of adjusting external initiatives to the city's own technological environment, with no real possibility of being involved in organizational and financial planning, as they merely respond to an 'external' stimulus. The reason behind this choice is that the City had to abide by anti-trust regulations concerning the delivery of services, but this case shows that this does not always lead to more efficiency, even though it did affect cost-effectiveness (in the new competitive setting, Venis cut 50 per cent of their budget for the development of some projects in order to win a tender).

However, the main problems that the city had to face are related to the integration of front-office projects, especially regarding the issue of smart cards for citizens and tourists. The illustration of this complex story needs a closer look at the background. Originally, the municipality had it in mind to solve the problems associated with the 'complex' use of the city by different categories of city users, through a technological device – a smart card. In general terms, this tool could allow the city to discriminate on price and information regarding the services delivered to different groups. Necessary technical preconditions were the 'chipping', which allowed the connection with the banking system, making this card useful for price discrimination and financially convenient; and the 'contactless' technology, making it possible to discriminate also on 'comfort' or 'time'. Different services could be charged on this card, for example, transport, access to cultural attractions, the use of toilets, etc., and moreover the owners of the cards could avoid queues. Citizens could use these cards to pay fees or use health services. It was important to keep tourists and other city users under the same 'technological' umbrella. The rationale is that tourists use the same services as citizens but they do not pay the taxes that cover the costs for the delivery of those services. Therefore, the card could be a tool to charge these additional costs. Moreover, different visitor groups (with different profiles) could be charged differently: for instance, those who book in the local hotels could be given an advantage – and thus an incentive – over the day trippers. However, by introducing a competitive element into the tourist market, these cards were also meant to attack the 'rent extraction mechanisms' that presently allow operators located in the proximity of the main cultural attraction to charge higher prices for goods that often are of inferior quality. In the 'virtual mall', all operators are the same and competition is genuinely on quality.

Thus, the city started to experiment with the CALYPSO card, as mentioned earlier. The city produced a large amount of cards and started to distribute them. At the moment of introducing this instrument, though, a number of issues were raised. The first regarded the target groups. Sectors of public opinion and of the local administration thought that the focus on visitors was wrong. The services, it was felt, should truly give an advantage to residents over visitors and in this way attack the Venetian 'tourist monoculture'. In the light of the previous discussion, this position was arguable. The issue is not getting rid of tourism, it is how to manage tourism and the rest of the economy in accordance with the city's aspirations, values and physical structure. This debate developed into a real institutional crisis. The 'visitor card' solution in the end prevailed and evolved into the *Venice Card*, which should be a contactless chipped card. Nevertheless, the city also decided to develop a different system; another card, *Carta di Venezia*, a sort of service card for citizens granting access to resident services, again experimenting in the EU context (LEADER project). A company was put up to develop this card, again as a partnership of more players such as the transport company.

The production of these cards, containing a chip and the electronic signature to access official documents, is expensive and slow (at present capacity, the city can deliver 150 cards). But there were other problems to come. In the same period the Italian Ministry for Innovation launched the project of a national ID card, utilizing a different technological platform, including an optic disc which, however, does not have enough memory to contain the digital signature. Then a 'National Service Card' was launched, with the same technical requirements, a different circuit and without an optic chip that basically replicated the *Carta di Venezia* format but relied on a different delivery network. The cards are bought by the cities who charge them with the services that are relevant to them and includes a digital signature. Some cities already have it but the validity is limited to the municipal territory. There is a problem of comparability between the actors who have registration authority. This version of the card now allows certification but there is no possibility of communication with Public Administration. The services enabled by the e-ID card are more limited, unless they adopt a 'light' signature standard, but massive investments have already been put into the 'strong' version of the digital signature. The Ministry imposed a standard on all the municipalities who had already progressed with similar projects.

The fundamental requirement for Venice is that any 'service card' should be contactless and distributed to tourists also. The challenge is to exploit the most interesting aspects of the two cards, integrate them and extend it to other facilities. The *National Service Card* is not yet started and has been actively pushed by many cities. The *Carta di Venezia* was started as a tool for transport and parking, to be integrated with health services: for instance, utilizing this card a physician can see the status of the patient and record new data. It could be extended to certification and access to municipal databases, protocol management with ID and password, with workstations in neighbourhood council offices and touch screens in newspaper kiosks. Other options include the booking of services of companies in the group for tourists.

As a matter of fact, there have been at least three portals concerning tourism in recent years: ALATA (focusing on the 'tourist region' of Venice), the APTs website (information on the city and its immediate surroundings, no online booking), Venice Card (selected information and booking of products in the city centre) and Urbis Limen (pre-booking of parking places, information on mobility in the whole Municipality of Venice). To these, two 'institutional' websites must be added, those of the Province and of the Region, providing extensive tourist information, statistics and services like webcams for skiers, thematic itineraries, etc. These initiatives have different 'status' (commercial v. non-commercial, private v. public or non-profit), regional scale (local v. regional) and range of services available (complete portal v. sector). They also developed autonomously one from the other – the only clear link is at the regional level with the APTs website being a part of the broader regional network.

Today the region is trying to improve the interactive character and the user-friendliness of its web-services, participating in an EU-Interreg partnership, Century 21, involving institutional and private partners like Oracle and Telecom Italy. The project aims to increase the availability of public services and improve quality of services. At this stage, the experiences of different countries are exchanged. Veneto focuses on tourism. The next step is to build a platform to integrate systems within the network. The project will extend the reach and develop some streams of the regional SIRT website (active since 1998), a tourist information system also made available to visitors through Internet. All the participants in the pilot stage are given €500 each to cover the infrastructure costs: 350 have agreed but only 250 have entered in the system, from a total of 3,200 hotels and 200 camping parks. In 2001, when contributions were no longer given, another 15 entered. Those who had already begun did not stop. This project utilizes a technological platform that is different from the one utilized by Venice Card. Although there could be several points of integration, the projects travelled on separate tracks. The Region sees little convenience in the Venice Card; they see it as being limited to a narrow range of services, with an unsatisfactory standard. Moreover, the impact on tourism is seen as limited to areas where it was easy to intervene anyway. The approach of ALATA is better thought of and, in fact, in the initial stage the Region was involved in the ALATA system as a data provider, but this was built as an association of 'local bodies' (Municipalities and Provinces) and the Region was not intended to be a key player. On the other hand, Venice Card has not tried to involve the region or the APT in its commercial plans or in the distribution system. These links, however, are necessary. APTs have the exposure and the expertise to undertake the distribution, and in fact they give information on Venice Card. However, being a 'public' portal, APTs are now prevented from delivering commercial services like online bookings. This hardly makes it competitive with respect to other private (but less community-oriented) websites. The change in the ownership regime will make the commercial move possible. However, APTs are keen to maintain a public orientation and their institutional role of 'promoters of territory' who care about quality aspects. The system will benefit from a convergence of strategies with many operators in the private sector, such as hotels and event organizers.

To summarize, in this host of initiatives many opportunities and investments have been lost. The only card that actually came into circulation is the *Venice Card* for tourists, which at the time of writing is not (yet) 'smart' and provides little system integration to achieve visitor management objectives. Whereas CALYPSO experimented for years and then a completely different standard was realized, *Venice Card* was introduced without any pilot experience and the 'business model' on which it stood is questionable. For one thing, it did not change the attitudes of the most 'problematic' visitors, as it does not provide a real incentive to come to Venice as an independent tourist staying longer than a day. This would only be a possible feature of the 'chipped' version of the card, which requires a sounder partnership structure between players involved like banks and tour operators.

Secondly, it was not sufficient to convince the cultural operators to become more entrepreneurial and cooperative. That is, the project especially encompassed the 'secondary' side of the tourism industry, but no real incentive for a restructuring of the cultural supply was offered with this instrument.

In particular, the points of integration and reciprocal self-reinforcement between the two main projects in which the municipality had a leading role were very promising. If ALATA had worked as a 'regional' infrastructure for the delivery of the Venice Card services (adopting the CALYPSO technology), the cultural cluster of Venice could have extended to the whole tourist region, realizing an innovative idea of 'multipolar cluster' which exploits the tourist attractions of a wide territory, further improving the efficiency in the organization of tourism. On the other hand, Venice Card could have comprised the whole set of services that were available under the ALATA partnership. This integration did not happen. The reasons are interesting as they hint at governance issues. A lesson learnt is that a leading regional actor should have been able to convince all the relevant levels of government to maintain the ALATA partnership. This requires a sounder approach in building a cohesive network of participants. If the metropolitan authority project is finally approved, this might be a decision-making platform to engage in this sort of negotiation with the 'outsiders' in the system.

5 Governing Access

The ICT access policy of the city can also be separated in two layers. The city is running its own 'alphabetization' programmes for civil officers and practitioners. Moreover, the city is trying to familiarize with ICT sectors of the population that could be particularly advantaged such as schoolchildren or the unemployed.

The city started to directly address the issue of ICT learning for officers when it discovered that its e-strategies needed increased capacity to handle procedures and content-building. In particular it was necessary to improve the 'acceptance' of ICT by the civil workers. People over 40 could easily feel 'condemned' as far as ICT skills are concerned and this could have hindered or stalled the capacity of the city institutions to improve their levels of service with regard, for instance, to citizens' support and visitor management. To this end, demonstration activities were utilized. The competent offices organized a survey (including a physical trip) to Bologna where e-government had been a reality for a decade. Venetian workers had the chance to see how user-friendly services could dramatically increase their efficiency and the quality of their work, with no significant loss of time or status. Today they have been exposed to the ICT revolution and can handle things easily. However, this issue highlighted that, because of learning constraints, the timetable for any significant improvement in terms of ICT adoption is a decade: the idea of achieving a structural change in three years does not make sense and plans have to be made for the long term. Other institutional actors are concerned with this issue. The Province has organized seminars on e-governance for its workers, which turned out to be relevant in a stage in which this institution was taking over important competencies in key sectors like tourism and employment. The Region, using the funds of the Century 21 project, improved the 'connectedness' of the Venetian hospitality sector, improving in this way the effectiveness of tourism management and information systems.

As regards the ICT education of local residents, the city organized its strategy utilizing the powerful institution of the District Municipalities, which are considered much more as a meeting place for the local communities than City Hall is. Seminars and schools were organized in the main municipalities working at full capacity under the recent decentralization framework, like Marghera. The aim is to provide low-cost or free access to families and social groups who normally would not be exposed to ICT (low-income families, unskilled workers, immigrants), improving the conditions for a successful transition of these communities to the 'knowledge society'. The programme extended to loans and grants through the school system and the Region to buy personal computers.

It should be noted, however, that the communities gravitating to the city are not limited to residents. First of all, of course, visitors are a group of city users with specific behaviours and values. They are also a very heterogeneous group. The city has an incentive to foster the creation of 'virtual communities' of potential visitors who could be familiarized with the city and its cultural offerings prior to their visit, dramatically improving their experience (and the impact on the city). Apart from the organization of contents, which we examined earlier, this strategy also needs to enhance access and IT familiarity. The latter is not to be taken for granted. Both among tourists and operators, the use of personal computers and similar equipment is not widespread, as was found in various studies (van der Borg et al., 1998; ICARE, 1997). Moreover, once in Venice visitors do not have the access to IT that they have at home and the reach of communication strategies is diminished. It is therefore necessary to organize access points, staffed or interactive, where visitors have access to information and databases and possibly can book their visits. This was ALATA's approach: however, after the dismantling of that system no real alternative was put in place. Another form of access is organized by museums, with computer rooms where visitors can retrieve information that may be a complement to their visit.

Other virtual communities can be stimulated among students, an important group of city users, and most importantly among patrons of the cultural supply of the city and possibly producers themselves. It should be noted in this respect that the Internet ownership and access facilities at the local academies is rather low. The local higher education institutions should coordinate with the city in widening students' IT access as a way of hooking them to the social capital of the city through knowledge-sharing and building of entrepreneurial capacity in fields that are directly relevant to the city's future, such as 'creative industries' and producer services.

6 Governing Infrastructure

The leading actor in the Italian infrastructure policy for communication has been, for years, the Italian telecommunications company Telecom Italia. The Italian environment for broadband differs in some significant ways from that in the other major European countries. The cable network in Italy covers only a few favoured areas so it cannot make a significant contribution to broadband. Italy also has high population densities in its cities, with a high proportion of people living in apartment blocks, and so offers a good opportunity for fibre-to-the-home solutions. This put Telecom Italia in a strong position to dominate the development of the broadband market in Italy, a tendency that the regulators have tried to restrain. TI's original launch of ADSL services was delayed from November 1999 to December 2000 by a judicial enquiry into complaints that TI was abusing its incumbent position.[4]

Venice had to be a first example of an 'ICT city'. The strategy involved infrastructure – Telecom Italy started a wide programme of cabling to provide fast access to the entire community, financed under the Italian Government's Socrates programme – for the institution of a 'brain' centre – the telecom centre S. Salvador – to provide inputs in terms of research and development to innovative content applications. This strategy was not developed to the full because of the oncoming changes in the telecom market. After deregulation, other companies became active. The city allowed new companies to put fibre-optics in the city, but this demanded a clear policy to prevent the necessary works from altering the physical structure of such a fragile city as Venice. The municipality was called on to have a central role to this respect, with a strong coordinating capacity of the host of actors that are involved in the system use, maintenance and financing. Today, four companies are allowed to put cables in the ground. Telecom Italia is also involved as the owner of the cable infrastructure and as a key player for content provision. It is sharing the use of its pipes with the municipality. One or two of the new pipes will be reserved for the use of the administration or other local institutions like the universities.

However, the new infrastructure is likely not to cover the whole city as in the original plans. The new regulatory system implies that only operations that guarantee a return are feasible and in this case cabling Venice may not be such good business in a short-term, market-geared calculus. The Socrates-funded cabling could be a good framework to work out of strict market considerations,

---

[4] In early 2001 competitive operators complained to the European Commission that Telecom Italia was being anti-competitive in unbundling and wholesale DSL pricing. Competing operators want a flat-rate wholesale HDSL charge, something the regulator AGC would like but says it cannot impose on Telecom Italia. Overall AGC has been more proactive in dealing with DSL disputes than most other European regulators and the Italian anti-trust authority fined Telecom Italia 115 billion Lire (approximately US$50m), for abusing its dominant position in DSL and broadband.

but the city needs a decision to revive this programme, which is a politically sensitive issue.

In the meantime, the city is looking at other opportunities. The Mayor supports the idea of developing an incubator of small and medium-sized 'creative' producers in the city centre, to be set up with structural funds, in collaboration with the VIU. A feasibility study is being drawn for 25,000 m², ICT-led development in a historical building in a popular area of the city centre.

7 Conclusions

The case of Venice offers interesting insight into how e-government might enable a complete change not only in the way in which citizens and institutions interact, but also in the very nature of the city and its development potential. In Venice, as in other cities, ICTs and the new economy that they support do not compete with traditional functions, but rather stimulate synergies at all levels. For instance, ICT can be used to compete on residential functions, allowing people to live in Venice though working in the network economy where the poor physical accessibility of the city is recouped 'virtually'. This vision should be supported by an e-strategy, at three levels: content building, access enabling, and infrastructure provision – in varying degrees, Venice has been active on all three fronts.

Venice's experience with tourism illustrates both the potential and the main obstacles of e-strategies for sustainable urban development. ICTs do indeed have the capacity of restructuring tourism in many dimensions. They can empower local operators and improve the network structure of the sector, reconnect tourism with the urban economy offering plenty of opportunities for a closer integration, enhance the value generation in the cultural sector, improve the capacity of visitors to make informed choices, improve the spatial distribution of tourist activity and allow a smarter, more selective marketing of the city's assets. These objectives, however, affect power relations in a highly contentious and fragmented system such as tourism. Governance theories offer an explanation of the difficulty of achieving social optima and regulation in a system in which informal players have a strong decisional power. Indeed, the realization of the city's interesting e-initiatives like the Venice Card, ALATA, the Carta di Venezia and other actions in the field of e-culture has been severely hampered.

First they had to challenge competing initiatives from different levels of government. They also met the partial resistance of private and most notably public actors, they did not involve all the relevant players and were not sufficiently backed up by the political class. This lack of cohesion can be explained in terms of a top-down approach – not considering the opposition of actors with conflicting objectives – and the lack of a 'business model' guaranteeing financial self-sustainability of these projects. Due to the public status of key players like APTs this model cannot be built without solid partnerships with the private sector, which have not been sought with sufficient consistency. Cultural and institutional factors are also a hindrance to the effective implementation of an e-strategy for tourism, as the scarce cohesion between levels of government with different political colour and the lack of an institutional level with exclusive competence on tourism matters – and, for that matter, on all metropolitan development issues – that covers the relevant regional scale. In the case of Venice, local projects were partly hindered by the concurrent launch of e-government national initiatives.

In the end, the general impression is that of lost opportunities, though there is still the possibility of reviving the original spirit of these initiatives in the light of a new stage of urban regeneration. At the time of writing, Venice is more than ever a 'divided city'. Any attempt to produce an alternative to the 'tourist rent', for instance with ICT, meets resistance. On the other hand, there are sectors of the city, not necessarily the original residents, for whom change is welcome. These communities are disconnected from the tourist business, for instance, students. For these groups – little represented in the political life of the city – Venice might one day become a place for the elaboration of issues. This population is made up of temporary citizens who might never be rooted in a qualitative way with the cultural texture of the city. Instead, Venice can promote its cultural heritage, both tangible and intangible, to attract the people who accept this lifestyle. On this ground they can compete with cities like Milan or Padua, where quality of life is now low, with high levels of traffic and criminality. These 'progressive' groups need to be empowered, given a voice and offered an easy way to reconfigure their role as a social and economic resource for the city. ICT and e-government can be powerful tools to achieve these goals.

For the future, a sounder approach in building a cohesive network of participants will be required. The political class itself needs to be thoroughly convinced of the potential offered by ICT to instil cultural and organizational cohesion among parties that are today hardly speaking to one another. The innovativeness offered by ICT produces little consensus in the short term, and this is a problem with short-horizon political cycles. If the metropolitan authority project is approved, this might be a leading decisional platform to engage in this sort of negotiation with the 'outsiders' in the system. Decentralization is also important. The city should allow bottom-up initiatives like the self-configuration of communities through ICT to flourish. The 'District Councils' which are being reorganized within the municipal administration offer an ideal meeting place and an infrastructure to both familiarize 'disconnected' groups with ICT and to recreate a network culture with access and understanding of the local cultural assets. In this way a cycle of development based on empowered e-entrepreneurs can develop. A real incubator can be organized in the media, design and cultural industries, generating precious knowledge and technical support for the development of an applied hi-tech vocation for the historical city.

The general lesson from this case study is that the networked, decentralized way of developing an integral e-strategy in a complex urban sector like tourism faces a lot of problems, not so much from the point of view of content, which can be easily harmonized, but rather from the point of view of the technical and regulatory conditions. From this point of view, a clear leader in the development process is fundamental. This policy-leader institution should be inclusive at the industry and spatial level, connected to but not dependent on political powers and financially sustainable. A Metropolitan Authority or Development Agency, for instance, could do this job, provided it worked in close association – possibly in a public–private partnership – with a telecom operator with the characteristics of Telecom Italy.[5] Policy development and implementation should be laid down in documents agreed on by the largest number of influential stakeholders, and be based on partnerships so as to ensure the widest possible technical and financial conformity.

References

Buhalis, D. (1997), 'Information Technology as a Strategic Tool for Economic, Social, Cultural and Environmental Benefits Enhancement of Tourism at Destination Regions', *Progress in Tourism and Hospitality Research* 3 (7), pp. 71–93.

Buhalis, D. and Licata, M.C. (2002), 'The Future eTourism Intermediaries', *Tourism Management* 23 (3), pp. 207–20.

Caserta, S., and Russo, A.P. (2002), 'More Means Worse. Asymmetric Information, Spatial Displacement and Sustainable Heritage Tourism', *Journal of Cultural Economics* 26 (4).

City of Venice (2000), 'Transforming Local Government with an IT Enabled Strategy – Venice's "Smart City" Strategy', Presentation.

City of Venice (2001a), *Towards an Economic Development Strategy for the City of Venice.*

City of Venice (2001b), *Venice's Economy: Current Trends and Future Prospects.*

Costa P. and Manente M. (1995), 'Venice and its Visitors: A Survey and a Model of Qualitative Choice', *Journal of Travel and Tourism Marketing* 4 (3), pp. 45–69.

Dente, B. (2001), 'Governing the Sustainable Development of Venice: Elements of the Institutional Planning Procedure', in I. Musu (ed.), *Sustainable Venice: Suggestions for the Future*, Dordrecht: Kluwer.

Go, F.M., Lee, R.M. and Russo, A.P. (2002), 'Bridging Global Divides: The Heritage Industries as Incubators of Change', paper presented at the International Conference *The Tourist-Historic City: Sharing Culture for the Future*, 17–20 March, Bruges, Belgium.

ICARE (International Center for Art Economics) (1997), 'Applicazione della Telematica alla Gestione dei Flussi di Visitatori', Research Center Telecom Italia S. Salvador, mimeo.

Klaassen, L.H. and van den Berg, L. (1989), *The City: Engine behind Economic Recovery*, Aldershot: Ashgate.

---

5 One such organization is Telepolis, developed by the City of Antwerp in partnership with Belgacom, Telnet, the city's main higher education institutes, the Chamber of Commerce and the urban region's Planning Department. It is an ICT network engaged in all aspects of urban management offering a successful template to similar initiatives in other European cities (http://www.telepolis.be/).

Rullani, E. and Micelli, S. (2001), 'Immaterial Production in Venice: Towards a Post-Fordist Economy', in I. Musu (ed.), *Sustainable Venice: Suggestions for the Future*, Kluwer: Dordrecht.

Russo, A.P. (2002), 'The "Vicious Circle" of Tourism Development in Heritage Destinations', *Annals of Tourism Research* 29 (1), pp. 165–82.

Russo, A.P. (2002), 'The Sustainable Development of Heritage Cities and their Regions: Analysis, Policy, Governance', unpublished thesis.

Van der Borg, J. (1991), *Tourism and Urban Development*, Thesis Publishers, Amsterdam.

Van der Borg, J., Russo, A.P. and Minghetti, V. (1998), 'The Application of ITT to the Italian Tourism Industry', Working Paper CISET No. 02/98, University of Ca' Foscari, Venice.

Interview Partners

V. Baldassi, Venis Informatica, Director.

A. Bolognin, Azienda Servizi Mobilità – ASM SpA, Director.

D. Bovo, Consorzio Venezia Ricerche, Director.

M. Ceselin, Information Systems Department, City Council, Director.

Mr Favaretto, Region Veneto.

S. Micelli, TeDIS Center, Director, Venice International University.

M. Milani, ALATA.

M. Munari, Region Veneto.

A. Povolato, ALATA.

G. Romanelli, Municipal Museums of Venice, Director.

Ms Spagnol, APT Tourism Promotion Venice.

R. Turetta, Neighbourhood Council Marghera, President.

J. van der Borg, Venice Card Ltd., President.

Chapter 11

Synthesis:
Comparing E-Governance in Cities

1 Introduction

In this final chapter, we present a synthesis of the findings of the various case studies. In section 2 we briefly summarize our framework of reference. In section 3 we compare the different urban e-visions and strategies that we encountered in the case cities. In sections 4, 5 and 6 we synthesize the experiences in the cities in the fields of content, access, and infrastructure, respectively.

2 Recalling the Frame of Reference

In Chapter 2 we developed a framework of reference in which we make a distinction between three 'local aspects' of ICTs: content, access and infrastructure (see Figure 11.1).

There are strong indications that the three local manifestations of the information society are interdependent and sometimes mutually reinforcing. We suggest that its dynamics can be represented as a local 'digital flywheel', which functions as follows. If there are more ICT users (access) in a city, it becomes more interesting for companies or any other actors to develop new services (content). The other way around, more (or better) electronic services (content) may increase the number of local users. If there are better online products or services available, the Internet becomes more useful, and more people are likely use it. This interdependence between access and content is well known in the economic literature on technology adoption (see Leighton, 2001). We have argued that 'turning the flywheel on' could bring benefits for cities in several respects: see Figure 11.2. Improved electronic services imply a higher quality of life for inhabitants: they have better access to improved amenities. E-government services may save public spending and reduce local taxes to the benefit of citizens and/or firms. Also, e-government may improve local decision-making as it improves the quality of management information. The quality of local electronic infrastructure is a factor of growing importance to attract or retain inhabitants (Healey and Baker, 2001). Wired homes have the potential of being seen as more up-market and desirable than others (Baines, 2002). Virtual communities can contribute to safety, social cohesion and political participation (van Winden, 2001). High-

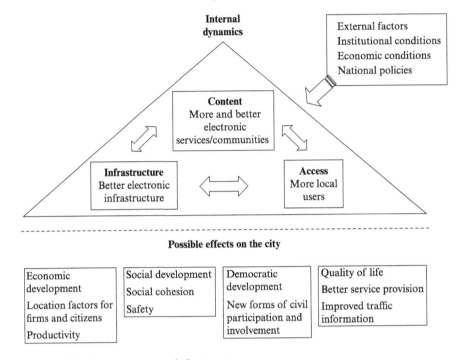

Figure 11.1 **Access, content, infrastructure**

Figure 11.2 **ICT policy, urban goals and organizing capacity**

quality ICT infrastructure is also important to attract to or retain firms in the region. Furthermore, ICT policies may bring 'first mover advantages'. If a region manages to create early mass in users and infrastructure, local firms may build an innovative edge. Especially, early critical mass of users may attract innovative companies and people into the city. The system takes off when a critical mass of users is reached.

A question that comes to mind is how local is the 'local flywheel' really? Clearly, its engine is not solely fuelled by local factors. External factors play an important role, too. In the first place, national institutional conditions matter. Our South African case study suggests that it makes a big difference whether the telecom market is liberalized and competitive or not. In addition, all other kinds of legislation influence the flywheel as well: for instance, electronic privacy and security legislation. Second, general economic conditions play a role. ICT use is strongly related to economic development levels. Richer countries and cities tend to have higher levels of ICT access, more content to offer and a higher quality of ICT infrastructure. Third, national policies can strongly influence the different parts of the flywheel. Regarding access, many countries have nationwide programmes for ICT in education or access policies for disadvantaged communities. In the field of content, national policies may encourage cities or other public entities to develop e-strategies and thus speed up the quantity and quality of content offered. With regard to ICT infrastructure, national governments can also deploy policies. For instance, the Swedish government has a policy that aims to construct fibre-optic networks throughout the whole country. Despite all this, our study has revealed that there is still sufficient scope for urban policymakers to do something.

This is where 'organizing capacity' enters the picture. The policy question we wanted to answer in our study is how urban policymakers can make the wheel go round. In our analysis, we applied a number of critical factors that play a determining role, which can be summarized under the heading of 'organizing capacity' (van den Berg, Braun and van der Meer, 1997). The first factor is the presence of a vision on e-governance and the translation of that vision into effective strategies. A vision helps to set a longer-term perspective, guide investments and link ICT initiatives with broader urban issues. The second factor concerns the presence of leadership. Previous research suggests that for any type of policy or strategy, awareness and active support of senior levels in the organization is critical. This also holds for e-governance strategies. Thirdly, the quality of local networks is an important determinant in the formulation and execution of e-governance strategies. Each dimension of local e-governance involves several partners: typically, these are local governments, citizens and technology suppliers, but other parties may also be involved such as other public agencies (public companies, financial service firms or other content providers). The different dimensions of e-governance (content, access and infrastructure) may require different approaches of network organizations. However, in general, cooperation is critical for the success of e-government policies. Therefore, in our analysis, we emphasized the pressing issue of public–private and public–public cooperation in the design and implementation of e-governance: how to involve the private sector in public e-government policies? How to align bottom-up initiatives in the city with general visions and strategies on the urban level? How can public agencies (within the city, but also on the national and international levels) cooperate in various ICT-related policy fields? Thus, networks are not only a means to get things done with urban stakeholders, but can also be a powerful tool to influence the 'external factors'.

3   Comparing Urban E-Visions and Strategies

In this section, we describe and analyse the various urban e-strategies and visions that we encountered in our study.

Most cities started to develop ICT strategies during the 1990s. These strategies were characterized by 1) a focus on technological issues, and 2) an internal orientation (how to use technology inside the municipal organization). Cities worked hard to create internal networks, and increase the use of personal computers and other ICT applications among their workforce, and to automate certain administrative procedures.

Urban e-strategies have evolved from an internal and technology orientation to a more outward-looking approach in which the focus is on the way ICTs can benefit the urban economy and society.

In the course of the 1990s, when the Internet emerged as a powerful and widely adopted new medium, cities started to rethink their role in the information society. New information technologies, especially the Internet, were increasingly regarded as crucial changers of urban societies and economies. This shifted the attention of urban policymakers increasingly outwards. Many cities started to worry about the emerging 'digital divide' between people with and those without access to ICTs. Also, cities became aware of Internet's new opportunities to have interaction with citizens, and other stakeholders. In the late 1990s/early 2000s, cities started to develop new ICT strategies that were more outward-oriented. These strategies primarily dealt with issues related to the question how ICT policy can contribute to urban development. They were less focused on technology and more integrative. This is reflected in the policy terminology: some cities no longer spoke of an e-strategy but rather an 'e-vision' (The Hague, for instance). In the wording of our frame of reference, content and access became key elements of urban e-strategies. Also, many cities began to realize that the benefits of ICTs could only be reaped if the introduction of technologies was accompanied by new ways of working, both within the municipal organization as well as in networks with other urban stakeholders.

Despite similarities and common patterns, in our case studies, we found substantial variety in the orientation of e-strategies.

The e-visions and strategies of our case studies reflect the cities' specific ambitions, problems and opportunities.

Our case studies make clear that the ambitions of cities are different, and reflect the specific problems and opportunities of those cities. Eindhoven and Tampere primarily consider their ICT policy as a way to strengthen their technological front-running profile. The Finnish city of Tampere is a prime example of a city with a very outward-looking and integrative e-vision. In the late 1990s, it developed its 'e-Tampere' strategy as a common umbrella and 'brand name' for its highly ambitious set of ICT initiatives. This strategy covers the business sector (aiming to develop a competitive ICT sector), the knowledge infrastructure (aiming to create excellence in ICT-related education and research), as well as the use of ICTs internally within the municipality. In Eindhoven, the equally ambitious and innovative e-city project reflects the city's drive to be 'leading in technology' (which is the city's slogan). In Venice, as expected, tourism-related initiatives are centre stage. The idea of the ALATA project, for instance, was to spread tourism more equally in the region by informing tourists about interesting attractions outside Venice. The e-culture strategy of the city aims to help cultural institutions in the city make better use of ICTs. In the relatively 'divided' cities of Manchester and Cape Town, ICT policies focused much more on transferring the benefits of technologies to less advantaged groups and neighbourhoods. This holds for content, access and infrastructure policies. Manchester's 'Eastserve' project in a deprived neighbourhood is a good example, where the city among others supports the creation of useful web applications for people in that area.

The Role of Leadership

In some of our case-study cities we found that top-level politicians were very committed to ICT strategies. Probably the most outstanding examples are Barcelona and The Hague. In Barcelona, the mayor and one alderman are both strongly convinced of the strategic importance of adequate ICT strategies and policies, and have put it high on the political agenda. The mayor discusses the content of the city's website with a number of people on strategic positions. In The Hague the mayor is also personally very committed to ICT and, in cooperation with the responsible alderman, has directed substantial budgets to e-government initiatives. This has certainly contributed to the city's front-running position in The Netherlands with regard to e-government. In Cape Town, we found that officials in the municipality are used to working in and experiencing an ever-changing environment. They are less 'entrenched' than people in many European cities where the dynamics and problems are less obvious. This attitude greatly facilitates the introduction of e-government, as this requires a great deal of internal adaptation as well.

4   Comparing Local Content Policies

In this section, we discuss the governance of local content. We defined local content as electronically available information, interactive services or other web content related to or concerned with a specific locality. Examples of local content are the local newspaper on the Internet, websites on the traffic situation in the city, information about events in the city, or the electronic services that the local administration offers to its citizens. It also includes the websites of firms or institutions that primarily serve a local market, such as community organizations, education institutes and non-profit organizations. Finally, it includes local virtual communities such as self-help groups, newsgroups etc. In this section, we will discuss the role of local governments in local content.

Providing Public Services Online

Electronic service provision is one of the key 'content-related' roles of local governments. In a number of cities, it is now possible to submit online forms, for instance for permits or allowances, or process other routines by electronic means. Our case studies vary considerably in the number and quality of online services they offer.

But within city administrations too, many differences exist. In Manchester, for instance, the housing department is the leader of the pack. On the Manchester city council website the Housing Department offers a 'home finder programme', which enables users to search for and subscribe to houses, after selecting a number of search criteria. Tenants can report necessary repairs online and book a date for repairs. The system links up with suppliers' systems and agendas (plumbers, carpenters, etc.). In The Hague, the local tax department has the most advanced services. Its website allows owners and users of real estate (citizens and companies) to check the value of their objects, to obtain the taxation reports and eventually to respond electronically. In Venice, the city is busy developing a city portal to give information and services to citizens, following the general directives of the Ministry for Innovation. These directives identify the services of the state and local authorities that can be managed through these portals, like online payments of taxes and fees.

Although electronic service delivery has the potential to reduce costs and increase productivity, it brings additional costs, because the other communication channels are kept open as well.

Cities see the Internet as yet another channel to provide municipal services. They keep the traditional channels (counters, call centres, information desks, etc.) open as well, since many people do not have an Internet connection and because large groups of people prefer to deal with the municipality in the traditional way. This means that the Internet brings along additional costs for cities, although it has the potential to increase productivity and save money.

As providers of electronic services, cities do not score well compared to large consumer-oriented companies.

When we compare the electronic service provision of cities with that of some (larger) private companies in the consumer market, the performance of cities is rather poor. Little of the potential of the new technologies has been realized. Truly demand-oriented services are still few and far between and the quality of city websites is often poor compared to larger companies in the consumer market. This relative backwardness can be explained by a lack of market incentives and the high quality requirements of public services in terms of privacy protection, identification and security. In all our case cities, national governments play a key role in setting legislation and standards in the fields of electronic identification, safety, responsibility, protection of data, etc. Some of our cities (Venice, Manchester and The Hague) feel frustrated by the poor progress of their national governments in these fields, which hampers their development of interactive services. Interestingly, the same national governments are increasingly providing incentives to cities to speed up e-government. In the UK, the national Labour government wanted to have all public services online by 2005 (DTLR, 2002). It encouraged cities to do the same (in partnership models) and gave them funding to do so. In competitive bidding processes, cities have to compete for national funds. In The Netherlands, three cities were appointed 'superpilots'. They receive substantial national funds to experiment with innovative e-services.

Encouraging Other Actors to Offer Services Online

Cities do not only produce local content themselves but also, in order to increase the quality and quantity of local content, many cities have taken measures to encourage local actors (citizens, firms, voluntary organizations and others) to create content. We found that they have several reasons for this.

- *To promote local culture and tourism.* Venice encourages the municipal cultural institutes to use ICT. In the short term, people will have access to the catalogues of the municipal libraries and municipal museums. The latter have international relevance, so the 'user group' is particularly large and interesting.
- *To help voluntary organizations that lack the means and expertise to use ICTs.* Manchester, for instance, supports an organization that helps local community and volunteers organizations to design and publish websites.
- *To improve productivity and competitiveness.* From this perspective, The Hague supports SMEs implementing e-commerce solutions, by giving them free consultancy days. Cape Town has established learning centres in public

libraries, where small companies in disadvantaged communities are helped to use new technologies. Eindhoven has a policy to promote the creation of broadband content by local companies.

- *To promote social inclusion.* The Hague actively promotes the formation of online communities. A website, Residentie.net, was initiated by the local government, as a platform for the creation of digital communities. Citizens can create thematic 'squares', for instance, a square for your own physical neighbourhood. Despite the city's ambitions with the project, visitor numbers are low and the formation of 'digital squares' falls short of expectation; participation in online debates is modest. The city of Eindhoven wants to set up an online community of broadband users. Remarkably, private companies are interested in participating. The Rabobank, one of the largest Dutch banks, is one of them. It views cyberspace as a new domain where local companies and citizens interact in new ways. The 'e-city', with its envisioned 84,000 inhabitants always online with broadband, could yield important lessons for the bank. It could show how interactions and transactions among citizens and companies may change in the future. For instance, banking services could become an integrated part of local e-commerce.

Encourage Online Participation and E-Democracy

Many cities see the Internet as a new means to communicate with citizens. 'E-democracy' initiatives have been taken by all our city cases. On a basic level, most of our cities put council decisions and policy documents online. Eindhoven is the most active city in our sample: it even broadcasts its council sessions live on the Internet. Another option for cities is to use the Internet to involve citizens in decision-making processes. However, this proves to be a difficult trajectory. The Hague has used its Residentie.net platform to organize online discussions on urban renewal projects and, recently, on youth policy. The city had a hard job getting people involved. In Barcelona and Venice, e-democracy clearly has less priority.

Content Organization and Web Strategies

In the early days of the Internet cities had no structured views on how to use it for their purposes. Cities' and municipal departments' Internet presence depended on the efforts of individual enthusiasts who put content online. Urban virtual communities ('digital cities') – in which citizens started discussion groups and undertook all kinds of experiments – emerged spontaneously. An example of such a community is the 'Digitale Hofstad' in The Hague, which still exists.

In the course of the late 1990s, urban managers became more aware of the power of the Internet as a means to inform stakeholders (citizens, tourists and companies), to communicate with them and to deliver services. In the first stages, Internet activities of cities remained very fragmented and strongly supply oriented. Typically, each municipal department set up its own website regardless of what the others did. These websites used to provide information only (no interactive services) and mainly reproduced existing information which was also available in folders, brochures, etc. There was very little coordination on the city level. This made it difficult for web surfers to find the right information.

In their presence on the web, cities are moving from a supply-oriented to a demand-oriented approach, but they do so in different forms.

At the time of writing (2004), most cities deal much more strategically with the Internet. They regard it as a strategic tool for city marketing, communication and service delivery. Many cities have developed an Internet strategy and set up structures to maintain the technical and content aspects of their Internet presence. However, our study reveals considerable differences in the way cities present themselves on the Internet. Portals are now the most common way to organize information in a more harmonized way, but they come in different forms. We discovered four types of portals: territorial, target-group oriented, project/topic oriented and municipal.

Territorial portals
Many cities have developed territorial portals to offer an overview of links to all kinds of websites in a specific area. Cities follow different approaches concerning the geographical focus of their main websites. Evidently, most cities have portals for the territory of the city itself. In Venice, however, the ambition is to develop its website as a territorial portal, a one-stop window for municipal services and firms operating in the region. This portal will give access to services like patents for starters, new shops, building permits, health services, fire brigades, etc. The idea behind this is that Venice is part of a functional region where many mutual relations exist. Barcelona follows a similar path. Together with the region of Catalonia and a number of municipalities, the city aims to create a single e-government portal for all municipalities in the Catalan region. In South Africa, the province of Western Cape (of which Cape Town is the capital) has initiated a provincial portal website Cape Online, which connects 450 websites of public organizations.

Other cities view the borough or neighbourhood rather than the region as the appropriate scale to organize a web portal. In the deprived area of East Manchester, a portal was developed enabling people to access services of seven agencies, among which the City Council, health and education providers, the police, and voluntary sector bodies. The underlying vision is to 'deliver online a wide variety of services to local residents in a manner that is appropriate to their situation and need, and which gathers services to suit individual customers and citizens'. The argument is that each neighbourhood has specific needs and problems, and should therefore be 'serviced' by a different portal.

Portals aimed at specific target groups

Many cities have separate portals aimed at specific target groups, which can be tourists, foreign investors or citizens. The city of The Hague has a separate website (in English) for foreign investors, on which it offers all kinds of information that could be relevant for corporate decision makers. The Hague also has a web portal for residents (Residentie.net) where they find information on local events, but also create their own 'community' content. Manchester has a portal for local voluntary organizations. Some cities (Manchester, Venice, Eindhoven) have specific portals for tourists, often maintained by the local tourist board and for traffic information. Some cities have competing private websites for tourist information (Venice, Eindhoven, Manchester). Remarkably, the city of Cape Town, with its large tourist potential, has no tourism portal.

Portals for specific topics or urban projects

Cities create portals, like websites for large events, projects or other topics. One example is the e-city portal that informs and communicates with potentially 84,000 people who are part of the 'e-city' project in Eindhoven.

Local government portals

All the case cities have portal websites in which they present local government information and public services. These portals contain links to the various departments, and quite often also to the types of portals described above. Beyond these portals, however, the content organization is still supply-oriented in most cases. The individual city departments are responsible for the maintenance of their own websites, on which they offer their own services and information. Typically, city departments have agreed on a common style and standard for their presentation. The Dutch case cities (Eindhoven and The Hague) have sites that are rich in terms of documents, reports, council decisions, etc. Some cities are beginning to make the shift from a supply-orientation (where individual departments of city administrations offer their content in an online form) to demand oriented and integral solutions. In this way, they make better use of the added value the Internet offers. The Hague, for instance, is implementing electronic services from a life-events perspective. A site has been designed for people who want to get married showing all the localities in the city where the marriage can take place and allowing for online reservation. Eventually, private companies may take part in this initiative as well. This example suggests that the Internet may become a trigger for new kinds of public–private cooperations in service delivery.

For cities, the Internet may be a trigger for new kinds of public–private partnerships in services delivery and information provision.

Who is responsible for the websites/portals?

In all our case cities, the question is raised of who should maintain the portals and who carries the responsibility for the websites. In the early days of the Internet, the technical people from the information technology departments were typically in charge. This stemmed from the view that an Internet site was primarily a technological issue which you needed experts to maintain. In the city of Cape Town, the central IT department is responsible for the coordination of the web development of the municipal organization. In many other cities, the communication department is in charge of the web portal. This reflects the idea that the Internet site is a communication channel rather than a technical issue. In many cases, an urban communications department collects information from the other departments or sets standards for the layout of departmental websites. In the last few years, cities have been introducing 'content-management systems' which enable departments to publish their own web content more easily. The city of The Hague has even outsourced one of its portals (Residentie.net) to a private telecom supplier. In Barcelona, website policy has a high priority. The mayor personally evaluates the municipal website every week.

Most cities have many web initiatives, but lack an overall vision/strategy on what they want to use the Internet for.

With the increasing number of interactive services offered on websites, the Internet is changing again: from a communication tool to an integrated services engine. This implies that Internet activities become more intricately linked to the entire municipal organization. The implications will be discussed in the next section. After our survey, we conclude that cities lack an overall integrated view on what they want with the Internet. Web strategies exist, but they are fragmented.

Re-organizing the Back Office

Implementing integrated services asks for radical changes within city administrations. In general, the Internet is a trigger to change the municipal organization and make it more client-oriented. This requires, among many other things, an integration of internal IT systems within the municipal organization.

The Internet works as a catalyst to make municipalities more client-oriented.

In all our case-study cities, every city department used to have its own ICT applications and data formats, could deal with its own 'technology supplier' and sometimes ran its own electronic infrastructure. However, to enable e-government, the different systems and departments increasingly have to communicate with each other. E-government implies that information has to be shared and exchanged by various departments. This holds for basic information on the population (names, addresses, etc.), but also for geographic information.

In many cities, the need for systems integration creates tensions between central IT departments and other municipal departments.

Among other things, the need for integration creates tensions between the central IT department and the IT units of individudal departments. Typically, the departments want to keep their freedom to buy what they want, saying that each department has specific requirements and should not depend on the 'monopoly' of the central IT department, while the central IT department argues that it can achieve cost savings (by centrally purchasing equipment, ICT infrastructure and support) and make sure that the various systems are integrated.

Different cities handle the tension in different ways. Barcelona and Venice have developed models that allow for system integration through some form of central coordination. In Barcelona, the city government decided that e-government required a centralized ICT strategy and centrally-managed facilities. The municipal Institute for Informatics (IMI: an autonomous institute, 100 per cent owned by the municipality) is responsible for this. Although municipal departments are free to choose other IT service providers, IMI has a *de facto* monopoly, because almost every IT system and application must fit into the highly integrated infrastructure of equipment, software and information. IMI has to negotiate contracts with each of the municipal departments. These are complex negotiations, because the partners have to find a balance of the clients' priorities – as represented by the department managers – and the global interests of the city – as represented by the director of the General Services Department. The city of Venice has founded Venis Ltd, a public company in charge of integrating the various information systems utilized by the municipal administration and delivering services. Until a recently they were the only provider, with total responsibility for the city's ICT services. Now they are fully responsible only for system management and integration, and compete with other providers for new systems and projects. In this transition, Venis had to change its mission. Now they do a not-for-profit job of adjusting external initiatives to the city's own technological environment, without a real possibility of doing some organizational and financial planning, as they merely respond to an 'external' stimulus.

For e-services, some form of central control over IT is indispensable, to endorse standardization and facilitate information exchange. This often leads to conflicts with departmental IT units that don't want to give up independence.

In The Hague there is much less central control. All of the 14 departments have their own information technology units, which are fairly independent. They operate the main information systems applications as well as the personal computers, the Local Area Networks and the midrange computer systems. After the outsourcing of the municipal computer centre in the early 1990s, maintenance of the technology and the information systems, as well as mainframe operations, were outsourced to a preferred supplier for a period of five years. After this outsourcing contract expired the individual departments outsourced these tasks in a rather uncoordinated way. There is no information systems integration. For e-government to be effective and efficient however, system integration is an important precondition. The city is searching for a partner to integrate its various information systems in a comprehensive way.

In Cape Town, the fragmentation of information systems was enormous. The current 'Unicity' of Cape Town was established in 2000 as a consolidation of seven municipalities. In the early 1990s, the number of municipal authorities was even greater. This legacy is reflected in the current IT context in the city. In terms of IT, the city finds itself with a multitude of IT systems: every municipality, and almost every department, has its own hardware, networks and applications. Some estimate that a total of 270 (!) systems are currently in place. This situation seriously hampers the management of the city, as the systems communicate with each other with difficulty, so the information provision does not meet the demands of even basic management standards. Both the city manager and the newly created centralized IT department were strongly convinced that a new, integrated information management system needed to be put in place in order to streamline information flows and integrate all the different systems. It was decided to put an integrative Enterprise Resource Planning (ERP) system in place, which would meet these requirements, in cooperation with a consultancy firm. The system will integrate human resources and payroll systems, financial accounting and management accounting, real estate management, revenue management, customer care and materials management/procurement, and will replace or integrate at least 150 of the current systems.

In Figure 11.3 the increasing need for organizational change within municipalities is shown: this increased need is a consequence of the development

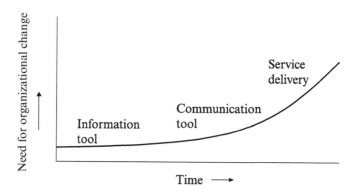

Figure 11.3 The increasing need for organizational change

of web content from a relatively simple information tool to a more complicated and integrative service delivery mechanism.

Strategic partnerships
For the integration of services, a reorganization of work processes within city administrations is a precondition. Generally, cities lack the knowledge and resources to manage these change processes. Therefore, they tend to engage in strategic partnerships with consultancy firms that combine expertise on ICT implementation/integration with that on organizational restructuring. The Hague works with Microsoft as key partner, Manchester with a consortium of ICL/Fujitsu and Deloitte&Touche, and Cape Town with Accenture/SAP.

In Manchester, the City Council was aware that implementing e-government requires both new organizational concepts and integration of IT systems, and that for such a complex operation, specific know-how was needed that was not available within the organization. Therefore, in the year 2000 the city advertised for a partnership with private business to improve local services. Consortia could bid for the job. Several bids came in and ultimately the combination of ICL/Fujitsu (a computer company) and Deloitte&Touche Consulting won. They proposed combining a business process re-engineering implementation with a re-arrangement of IT systems. In the negotiations that followed both parties found it hard to give substance to a real 'strategic partnership' between city and the businesses. Ultimately, cooperation took the form of a risk-reward strategy: that means that the consortium will be rewarded when results are good, but it also carries part of the risk of failure. The time span of the cooperation will be between seven and 10 years. This is typical for ICT investments after the collapse of the 'dotcom boom': cities are willing to invest in ICT but want to be sure that benefits in terms of costs savings or quality improvements will be achieved.

After the collapse of the dotcom boom, cities are still willing to invest in ICT, but they no longer give a free hand to the IT companies: they demand concrete results in terms of cost savings or quality improvements.

Our case studies reveal that cities have difficulties finding the right partnership model. Key issues are how to share risks and returns between the partners, how to keep control of the change process (this is not easy for cities, as their commercial partners tend to have the better knowledge base) and how to avoid lock-in into a particular system or supplier. In addition, the commercial consortia are not always aware of political and bureaucratic peculiarities and sensibilities of the municipal organization.

Outsourcing or Not?

System integration is an important precondition for e-government to be effective and efficient. Information and communication technologies play a key role in realizing and maintaining integrated services. However, many municipalities do not consider information technology as a core business and find it hard to operate online services 24 hours a day, seven days a week by themselves. Thus, in many cases they look for cooperation with the private sector, which often takes the form of *outsourcing*. The issue comes up every time additional resources have to be allocated to information and communication technologies and every time the city is in need of new ICT professional skills. Since good technological infrastructures tend to be hidden behind the screens and the organizational workflows, technology is often considered to be 'just a tool' – a cost centre rather than a key asset.

We believe that this issue of outsourcing should be approached with the utmost caution. As we have seen, the success of e-government in *Barcelona* relies heavily on a municipal information technology centre that ranks among the best of its kind in Europe. The Municipal Institute for Informatics (IMI) is an autonomous institute, 100 per cent owned by the municipality, and attached to the director of General Services. Under the leadership of IMI the information systems architecture of Barcelona have been developed over the years and has become a solid architecture since the early 1990s. Since that time, the Institute has been quite successful in matching legacy information systems with the current requirements of interactive and integrated applications. The city government of Barcelona decided that maintaining this centralized ICT strategy and centrally-managed facilities is a prerequisite for integrated attention to citizens as well as for territorial and functional decentralization.

Every year IMI has to negotiate contracts with each of the municipal departments. These are complex negotiations, because the partners have to find a balance between the clients' priorities as represented by the departmental managers and the general interests of the city as represented by the Director of the General Services Department. However, they share a common understanding of the public interest. In consequence, although municipal departments are free to choose other ICT service providers, IMI has a *de facto* monopoly.

Many cities in Europe lack such a strong municipal service provider. Although they may have a clear vision of and strategy on e-government, they find it more difficult to implement the strategy. They lack a history of technological experience as well as the direct access to – and control over – the application developers, the operators and the engineers that *make it work*. In fact, many of those who outsourced their IT departments in the past regret doing so. The city of Leipzig – not one of our case studies – even bought back its IT department a few years after it had been outsourced and modelled its new position in Barcelona's municipal structure.

The city of The Hague was already outsourcing its municipal computer centre in the early 1990s. Maintenance of the information systems, as well as mainframe operation, were contracted out to a preferred supplier for a period of five years. Only the telecommunications infrastructure (speech, data and mobile) remained under the responsibility of a dedicated central unit called *Haagnet*, which was created at the same time. Under the conditions of the outsourcing contract the municipal departments were responsible for the Service Level Agreements for their own applications, personal computers, local area networks and mid-range computer systems. After the contract expired the departments chose to implement individual contracts with different technology partners in a rather uncoordinated way. As a consequence, when the Internet came up as a channel of service provision, the information systems were far from the level of systems integration that Barcelona had realized in the meantime.

Manchester City Council was aware that implementing e-government requires both new organizational concepts and integration of ICT systems. Such complex operations need specific know-how that was not available within the organization. The City Council chose to enter into a strategic partnership with ICL/Deloitte&Touche, which proposed a combination of business process re-engineering with re-arrangement of ICT systems. As a next step, the city did not just simply contract the job out but engaged in a strategic partnership in which the risks and benefits are distributed between the city and the consortium. Both parties have found it hard to give substance to a real 'strategic partnership' and the partnership arrangement did not preclude the Council from entering partnerships with other private sector suppliers as appropriate. In the process, the relationship between the consortium, the central ICT unit and the ICT units in the municipal departments has been troublesome at times. In some instances the departments have worked together with the consortium without consulting or involving the central ICT unit.

In South Africa 'the public option comes first'. Private sector involvement in public service delivery is an option only if the public sector cannot do the job well enough itself. As we have seen, in 2000 the city of *Johannesburg* chose the private option. As in Manchester, outsourcing the ICT services was the best way to acquire the technological expertise needed to develop responsive government and good quality public service delivery. The city's financial position was in poor shape, ICT management was underperforming, there were hundreds of ICT costs centres spread throughout the administration, the level of internal ICT services was poor and there was an apparent lack of skills for the delivery and use of desktop and LAN services. Outsourcing to the Sebedisana consortium was the quickest and the easiest way to obtain the required expertise. Meanwhile, the city retained responsibility for strategic choices.

Johannesburg has been struggling to implement its responsibility for strategic choices at an appropriate level. In the beginning the municipal ICT department was in no position to take up its role because of lack of skills, lack of mandate and lack of authority. It was quite difficult for the municipality to implement a corporate CIO unit that would address *ICT issues* from the *municipal core business* perspective, whilst being an expert in both fields – countervailing the expertise of Sebedisana. Perhaps the appointment in October 2002 of a Chief Operations Officer (the highest administrative position after the city manager) has been an improvement. Until then IBM South Africa was actually leading the way to e-government as Johannesburg's knowledge partner in the outsourcing contract.

In the meantime the business case has probably been saved by existing personal linkages. Before the outsourcing took place, many employees of Sebedisana used to work for the municipal departments and enterprises. They know the business and the people in the customer organization and they have acquired the new skills needed for e-government in the Sebedisana consortium.

Cape Town chose the public option. In 2000 the city found itself with a multitude of ICT systems, very much like Johannesburg. Every district and almost every department had its own hardware, networks and applications. This situation seriously hampered the management of the city. Both the city manager and the central ICT department were strongly convinced that a new, integrated information management system needed to be put in place in order to streamline information flows and integrate all the different systems. It was decided to put an integrative Enterprise Resource Planning (ERP) system in place, which would meet these requirements. The city decided a collaborative partnership with Accenture. As in Manchester, the process of implementing the system went hand-in-hand with a reorganization. The size and the impact of this project are substantial and the speed of implementation is impressive. It could only be generated by a strong ICT department, which not only managed to mobilize substantial internal and external resources to implement the system but also convinced the Council of the need to get things done. In this sense it resembles Barcelona. However, to keep control and coordination over all of the complicated ICT functions in the long run, the city of Cape Town may eventually have to engage into strategic partnerships with ICT companies. Cape Town stands firm, however, in that it will ensure its strategic control in any case.

The examples from Barcelona, The Hague, Manchester, Johannesburg and Cape Town certainly yield policy lessons for other cities that struggle with the problem of technological expertise for local government. Some of the main recommendations are:

- *The city should have a clear vision of the value of ICT.* It is not 'just a tool' or a cost centre but a key asset for municipal service provision, electronic government, workflow management and enterprise resources planning. ICT implementation has become a strategic issue. It touches the core business of the city and exerts substantial influence over the quality of service delivery.
- *The public option comes first.* However, many cities do not have all the high-level expertise needed in their own organization. They need to involve private partners and eventually outsource part or all of their ICT facilities. If the private option emerges, a high-level strategic partnership between the corporate unit and the private partners involved is required, keeping the city in the driver's seat.
- *Good working relations between the service provider(s) and the municipal departments should be encouraged.* The municipal departments have the best knowledge of the business of service delivery. However, a main threat to e-government and the integrated information architecture may arise when

departments act on their own and involve business consultants for supposedly departmental issues. In many cases these consultants expose a rather biased view of the departmental interest and their own scope of experience, paying too little attention to the corporate municipal perspective. In those cases where the ICT facilities have been outsourced they may lend too much of an ear to the outsourced unit and its mother company. Only a strong corporate level of ICT expertise in the municipal organization can deal with such threats.

• *The city should retain a majority say.* One of the issues at stake is that sooner or later the mother company of the outsourced ICT department always wants to impose its own professional standards for workflow, technology and information architecture. Furthermore, in the free market there is always a risk that ownership of the outsourced ICT department or its mother company will change. These issues should be properly addressed in the partnership model, the ownership and the influence over the ICT services enterprise.

• *The city needs an expert strategic unit for ICT at corporate level.* Such a unit should be powerful by virtue of its political backing and influential by virtue of its professional authority towards the municipal departments.

All in all, outsourcing and other forms of strategic technology partnership are key issues for the delivery of e-government. Further research is needed to deepen our understanding of the opportunities and the risks that are at stake, as well as the strong points and the weaknesses of the different partnership and ownership models.

The Role of National Policies in Local Content

In our frame of analysis we assumed that national conditions and policies would play an important role in the governance of local content. In our case studies, we found substantial evidence to support this claim. We found that national governments have an influence on the production of local content in a number of ways. They play the following roles: 1) setting targets for local governments to put public services online; 2) financially supporting local initiatives; 3) facilitating knowledge exchange amongst localities; and 4) creating conditions for e-security and privacy.

1 *Setting targets.* An example is the UK, which set the target to put all government services online by 2005. It also made funds available for which cities have to compete. This helps to put e-government higher on the local agenda.

2 *Financial support for cities to implement electronic services.* In The Netherlands, the national government supports ambitious cities with respect to e-government through the 'Superpilots' initiative. The Hague and Eindhoven receive substantial funding (some €450,000) from this scheme. Likewise, in Germany, after a national competition in 1998, three cities were selected

out of 136 applicants to be pilots for e-government: Bremen, Esslingen and Nurnberg. Electronic identification is one of the key elements of the urban strategies. The National government invested a total of €25m. The Ministry of Economy and Technology was in charge of this programme. Total investment in all the projects in all the cities amounted to €60m.

3   *Facilitation of knowledge exchange of best practices.* Almost every country does this. In The Netherlands, local e-government initiatives are supported through the OL 2000 project. This project organization has created a 'catalogue' of public services that can easily be put online. In Germany, to transfer successful projects to other German cities much attention is paid to disseminating the results of 'model projects' in seminars, etc. From some evaluations, however, it has proved very difficult to 'transpose' experiences for the one city to the other, due to differences in specific organizational circumstances.

4   *E-security and safety.* All the countries studied are trying to develop e-security and identification standards on the national level. However, they tend to be slow in doing this, thereby sometimes frustrating progress in e-government on the local level. In The Netherlands, a 'public key infrastructure' is under development. Digital identity will be put on a smart card that will enable transactions with all public agencies. However, several public organizations (among which the national tax department) considered the progress too slow and introduced their own systems. Sometimes, there are conflicts between the national and the local level. In Italy, the Ministry for Innovation launched the project of a national ID card. However, it utilized a different technological platform than the 'Venice card' developed by the city of Venice. A 'National Service Card' was also launched. Compared to the Venice card, it had the same technical requirements. It basically replicated the Venice card format but relied on a different delivery network. The cards are bought by the cities who charge them with the services that are interesting for them, among which are digital signatures. Some cities already have the card, but the validity is limited to the municipal territory. The services enabled by the e-ID card are more limited, unless they adopt a 'light' signature standard, but massive investments have already been put in the 'strong' version of the digital signature. The Ministry imposed a standard on all the municipalities who had already progressed with similar projects.

In the field of e-security and identification, there are often conflicts between national governments and local governments.

Overall, we found that The Netherlands and the UK are the most active countries with regard to e-government introduction. In The Netherlands, a special minister was made responsible for ICT affairs and pushed e-government very hard. In the UK, Prime Minister Blair is a known supporter of proactive ICT policy. In our sample, Finland takes a middle position: it leaves the initiative to the local authorities. This is in line with the very decentralized organization in the country.

5 Comparing Access Policies

A very basic aspect of the local information society concerns the degree of access to technologies by the (various segments of) the urban population. Access to ICT has several dimensions. It includes not only the ownership of hardware devices but also the capabilities to use information technologies and access to the Internet (SCP, 2000; Mitchell, 1999).

The degree of ICT access (ownership, Internet access and e-literacy) among the population can be linked to several dimensions of urban development. First, levels of adoption are positively correlated to economic opportunities for cities. Second, we assume that ICT may help to tackle social problems and reduce societal inequalities in cities. Third, higher levels of ICT adoption may contribute to urban accessibility, and fourth, to the quality of life of individual citizens. These issues need some elaboration.

- *Economic development*. Higher levels of access bring more growth potential to the local economy. Urban regions are concentrations of ICT-intensive economic activities. Hall (1998) shows that many new urban jobs in the information economy require high levels of e-literacy. In several cities, there are severe shortages of ICT-skilled staff, hampering economic development. From this perspective, higher levels of adoption are desirable. Second, higher levels of Internet adoption by the urban population entail opportunities for local firms to introduce e-business. The more users in a city, the more interesting it becomes for (local) companies to offer online services. In 'high access cities', local companies may build a lead in e-commerce and strengthen their competitive position.
- *Social cohesion*. Higher levels of access to ICTs may contribute to ease social problems and reduce societal inequalities. Socially isolated or less mobile citizens can use the Internet to connect with others. Social cohesion in deprived neighbourhoods can be strengthened when people have a new means to exchange information and express their view on developments. It opens new ways to involve citizens in urban renewal policies. Better ICT capabilities of weaker social groups (unemployed, ethnic minorities, low-income groups, the elderly) may improve their chances on the labour market. This may entail positive economic side-effects for the urban economy.
- *Accessibility*. The Internet (or other ICT systems) offers the possibility to organize, process and distribute real-time traffic and public transport information. It could thus contribute to a more efficient use of the existing infrastructure and improve the quality of public transport. However, to a large extent the impact depends on the degree to which the population has access

to the technology. As long as only a fraction of the population has access to the information, the impact will remain small. From this point of view, raising Internet adoption rates is a weapon to fight congestion.

- *Quality of life.* ICT access can also improve the quality of urban life in several respects. For instance, the quality of health services can be improved enormously by introducing tele-consulting services: disabled people no longer have to travel long journeys to a distant hospital for a simple diagnosis. In leisure, the accessibility of events, concerts, attractions, etc. can be greatly improved if ICT is used to provide citizens with customized information and online reservation possibilities. It brings within reach a broader range of leisure services and greatly reduces search costs. Furthermore, higher access levels increase the scope for e-government, which can make life easier for citizens, improve decision-making processes and potentially reduce local taxes when e-government brings efficiency gains. However, whatever content is being developed, access levels determine to what extent the benefits are really capitalized. *The Economist* (Symonds, 2000) puts it even more strongly: 'for e-government to succeed fully, the dream of Internet access for all has to become a reality'.

In most European cities, still only a minority of the citizens has a computer at home or access to the Internet. Large groups lack basic ICT skills. Empirical studies confirm that weaker social groups particularly (unemployed, ethnic minorities, low-income groups, the elderly) show low levels of ICT adoption (SCP, 2000; DTI, 2000). People 'on the wrong side of the digital divide' lack access to information and services and do not benefit from the new possibilities, which reinforces their isolation and backwardness. Lack of ICT skills reduces their chances on the labour market. Many cities, including all our case cities, consider these low levels of ICT adoption as an undesirable situation, and try to speed up the adoption of ICTs. In this section, we describe what our case-cities are doing to fight this digital divide, and which urban actors they involve in their policy design and implementation.

Large Differences Exist in Levels of Access between the Various Cities

In our case cities, we have focused on the adoption levels of personal computers and Internet. We found the highest levels of access in Eindhoven and The Hague. The lowest levels are found in Cape Town. The levels of access to technology reflect the national situations. Within cities, large differences exist. They are greatest in the divided city of Cape Town, where entire slum areas do not even have fixed telephone lines. But in Manchester and The Hague too there is a substantial digital divide between rich and poor areas. Cities focus their efforts on helping to get people online, or develop skills to work with computers: different ways of doing this are summed up in Table 11.1.

- Some of our case cities have opened or supported special ICT centres, with the aim of helping groups with low levels of ICT adoption to make the move towards the information society. Table 11.1 shows a number of ICT adoption policies. Manchester has its 'electronic village halls'. They offer Internet terminals with free access to the Internet; also, several training and education programmes are offered at low fees. The Hague has similar ICT learning centres, based in public libraries. In Johannesburg, kiosks and other access provisions are to be implemented in public places, in particular in the multipurpose community centres of the townships. It would start with approximately five access points per community centre. Expectations are that it will evolve to a total of 1,000 access points across Johannesburg. As yet, however, many issues have to be resolved, such as user education, training of instructors, maintenance and supporting staff, communications infrastructure and the physical security of the equipment.
- Second, all our case cities put Internet terminals in public places (libraries or kiosks). This enables people who do not have a computer at home to access the Internet. It is a relatively low-cost type of policy. In the cases of Tampere and The Hague, these locations also offer training and assistance in the use computers and the Internet.
- Third, many cities allocate resources to improve computer and Internet availability at schools. In Johannesburg, the City Council aimed to facilitate access to computers for all schoolchildren before 2005. In Cape Town many schools are poorly endowed with ICT equipment. The city is investing in personal computers, as this is considered key to improving the chances of children in poor neighbourhoods. Some schools are opened at the weekends, to allow community groups to use the Internet and receive IT-related training. In Cape Town, some large IT companies contribute to access policy, from a social responsibility point of view. In a community access project, an Internet provider offers free Internet access and a large computer company finances part of the equipment. The city of Cape Town wants to make optimum use of private companies' support in its policies. Using GIS (geographical information systems), it is carefully 'mapping' the digital divide, to find out which areas are particularly poorly endowed with ICTs. The study will help the city to direct private initiatives into the areas where the needs are greatest.
- Fourth, some cities directly address the digital divide by offering ICT equipment at reduced prices. In our study, Manchester promotes individual ownership of devices and home access. In a particular area in East Manchester, citizens can buy a brand new Internet-ready computer with monitor and colour printer for just €317. In The Hague, families living on welfare can get free personal computers. Also, the city offers free Internet access to its citizens through the Residentie.net project.
- Fifth, a relatively new type of 'digital divide' policy is the promotion of broadband access: Manchester and Eindhoven are doing this. The City Council of Manchester is building a wireless broadband wide area network

Table 11.1 Access policies

'Governing access'	Barcelona	Cape Town	Eindhoven	Johannes-burg	Manchester	Tampere	The Hague	Venice
ICT centres for special groups	0	0	*	*	**	0	*	0
Internet terminals in public places	*	*	*	*	*	**	*	*
Personal computers/ Internet in schools	0	*	*	**	*	**	*	0
Reduced tariffs for ICT equipment/Internet access	0	0	0	0	*	0	*	0
Broadband access promotion	0	0	*	0	*	*	0	0
ICT training at reduced fee	0	*	*	*	*	*	*	*

0 = No policy
* = Modest policy efforts
** = Strong policy efforts

Note: The scores are only rough indications, based on a study of policy documents and interviews with key policymakers.

(WAN) in the deprived borough of East Manchester. This will enable people in the area (4,500 houses) to have free access the local intranet. Eindhoven is promoting broadband use as well. In a designated area of 84,000 citizens, the city offers a demand subsidy to broadband users, regardless of the type of broadband connection (cable, DSL, fibre-optic or wireless). With this policy it hopes to establish a critical mass of broadband users in the city, which will be an interesting test market for broadband content providers (see the previous section).

• Finally, all our city cases promote the development of various ICT skills by its citizens. In Tampere, the Internet bus 'Netti-Nysse' offers a mobile solution for access and training. This Internet bus stops at different business locations and residential areas on a regular basis. Interestingly, in The Hague, where courses are offered almost free in public libraries, there is a shift in demand from basic courses (how to use a computer, how to surf on the Internet) to more sophisticated skills, for instance, web design. This marks the maturity of access in this city.

As this survey shows, cities are quite active in promoting access to new technologies among weaker groups in the population, with Barcelona as a noticeable exception. In Barcelona, policymakers see no role for themselves in promoting access. In their view, commercial companies have already provided access for all: the large number of Internet cafés in the city offers sufficient opportunities for every citizen to connect to the Internet or to use a computer for a low fee.

There is still a debate going on whether (local) government should have a role in promoting ICT adoption. It can be argued that the adoption of new information and communications technologies follows a similar path to that of other high-tech products such as the video recorder or the television: the well-known S-shaped adoption curve (Leighton, 2001). When a new technology is introduced, at first the number of users is low and the prices are high. Next, with falling prices and improving user-friendliness of ICTs, the number of users/owners of the technology grows quickly. As time goes by, the market mechanism will provide for further adoption of ICT among people who derive value from it. Proponents of policy intervention, on the contrary, argue that ICT adoption can be of strategic importance for urban development in many respects. The returns of policy intervention to speed up adoption may, under some conditions, be high. What we can learn from the adoption literature is that efficient policies should preferably not be generic but targeted at non-adopting groups (see van den Berg and van Winden, 2002; van Winden 2001).

6 Comparing Electronic Infrastructure Policies

In this section, we review the role of cities in the provision of electronic infrastructure. We will see that from a rather passive attitude, cities are becoming much more active to promote the rollout of new electronic infrastructures.

Our research suggests that cities are becoming much more active to promote the rollout of new electronic infrastructures.

In this section, we use not only examples from our case cities, but other examples as well. First, we will discuss the availability of digital infrastructures. Next, we focus on emerging urban infrastructure policies. We then discuss the benefits of local/regional broadband policies, followed by the risks and costs. Finally we present some conclusions regarding ICT infrastructures.

Electronic Infrastructure in Space

The various types of electronic infrastructures (copper, coax lines, wireless networks, fibre-optic lines) can be regarded as the transportation system carrying the bits and bytes of the information society. The infrastructure landscape in cities has changed dramatically in the last decade. First, the number of electronic infrastructure networks has increased (several new mobile networks have been put in place in the last decade, but also high bandwidth fixed lines and satellite-based systems).

Spatial differences in infrastructure endowment have become wider, due to telecom markets liberalization and a declining prevalence of universal service obligations.

Second, the spatial differences in infrastructure endowment have become wider, due to telecom markets liberalization and a declining prevalence of universal service obligations. The quality and availability of electronic infrastructure differs both within and between cities. Typically, because of market size, larger cities are better endowed than smaller cities or rural areas, and within cities, richer neighbourhoods and business districts have better infrastructures than poor neighbourhoods. In this perspective, Graham (1998) notes the emergence of premium network spaces. These are very localized areas in large cities (like London's financial district) that have superior connections not only internally but also with similar places in other cities. For the location of business, particularly information-intensive service companies, the quality of broadband access is a major location factor (Healey and Baker, 2001). Broadband is different from 'narrowband' dial-up access in two important respects: first, it offers more capacity, and second, broadband technologies entail an 'always on' connection: the user does not need to dial into a network but is always online.

Copper telephone lines are almost ubiquitous, at least in Europe. This old technology is still the infrastructure for most of the Internet users. Telephone lines can even bring high-speed access, using various DSL (digital subscriber line) technologies. However, DSL is not available equally across space. Operators have to invest to make their networks ready for DSL and prefer to invest first or most in areas where the likely number of users is highest and/or costs of updating the lines are lowest. This has an impact on the spatial distribution of infrastructure and services. In The Netherlands and the UK, the supply of DSL is unevenly spread; in some neighbourhoods there is only one supplier while in others there are several, and in some neighbourhoods DSL is not available at all. The situation for cables – constructed for television but increasingly used for broadband Internet – is slightly different. In some countries (The Netherlands, for instance), the cable network has coverage of almost 100 per cent: cable Internet is available in all the major cities but not in several rural areas. In other countries (UK, France), there are considerable differences in coverage between large cities and rural areas. Within cities, some areas are cabled and others not. In Manchester, more well-to-do neighbourhoods have a cable network, but the poor borough of East Manchester has none. Mobile networks became available throughout the 1990s in all our case studies. In some areas of our case-study cities more people have a mobile phone than a fixed landline connection. In the near future, when high-capacity UMTS frequencies will be put into use, mobile networks are likely to become more important as conveyors of data. Another promising new technology, still in its testing stage, is the use of the electricity network for broadband data traffic. Speeds are comparable to DSL technology. Nuon, a Dutch energy company, is testing the technology in 180 households in the City of Arnhem. In Germany, the technology is being tested in 20 areas. By the end of 2001, 5,000 people were connected, 2,000 of them in the city of Mannheim (*Algemeen Dagblad*, 2002). As every household has electricity, this has the potential of bringing broadband within reach of every citizen.

In our South African cases, the infrastructure situation is inferior to the European one. Cape Town claims to be the 'best wired' city of Africa, but nevertheless, from an international perspective, the IT infrastructure is poor. This is mainly due to the monopoly position of Telkom, the incumbent telecom operator. Prices are relatively high and quality of services poor.

In several cities, grassroots initiatives have been taken to speed up the availability of broadband infrastructure. In Sweden, several social landlords are putting broadband infrastructure in their apartments. In Manchester, residents of a local authority housing estate have worked together to create an intranet, with fast and always on Internet access (Baines, 2002). In Amsterdam (but also in numerous US cities), individual broadband subscribers give each other access to their mobile home networks, thereby creating a citywide mobile broadband network. Throughout Europe there is now a debate going on about the role of local governments in providing broadband to its citizens. In the next section we will see how local governments are starting to play a stronger role in broadband provision.

A Typology of Urban Broadband Policies

In this section we will discuss and compare urban broadband policies. It draws not only from the case studies that are included in this book, but also from other cases identified in the MUTEIS project (Macro Economic and Urban Impacts of Europe's Information Society). These cases are Amsterdam, Groningen and Stockholm. Additionally, we used Internet resources to obtain additional information.

In our survey of European cities we encountered a number of local broadband policies. They can be divided in five types: 1) the formation of community networks; 2) generic supply policies to provide citizens and companies with broadband; 3) policies to increase broadband supply in newly built/redevelopment areas; 4) policies to bring broadband to specific groups in the city; and 5) policies aimed at boosting the demand for broadband. As well as these policies, cities can, of course, also choose to not have a broadband policy: in the e-governance research the cities of Cape Town, The Hague, Barcelona, Tampere, Johannesburg and Venice were found not having a specific broadband infrastructure policy.

Table 11.2 contains the typology of the policies, and summarizes each type's policy objective and rationale. Below, each type is described and exemplified.

Community networks
In many places in Europe and the US, community networks are being constructed. In one typical form of a community network, local governments create an fibre-optic infrastructure that links up public and semi-public organizations in the city and let service providers compete on that network.

Many local governments create an fibre-optic infrastructure that links up public and semi-public organizations in the city, and let service providers compete on that network.

The city of Groningen (located in the northern part of The Netherlands) is an example. Like any Dutch city, Groningen is endowed with copper and coax cable infrastructure. Every citizen and company has access to this. However, Groningen companies pay more for broadband connections than companies in several other towns, because of the long distance from Groningen to Amsterdam where the main Internet exchange is located. In Groningen, prices for glass fibre (leased lines) are five to 10 times higher than in the Randstad area (NOM, 2002). To improve the local broadband situation, the city has taken the initiative of creating a Community Broadband Network. In the long term, the Community Network plan aims to create a citywide public fibre network. In the first stage the goal is to connect around 170 public buildings in the city of Groningen to the network. The

Table 11.2 Typology of local broadband policies

Type of policy	Policy objective	Rationale for intervention
Community networks	Reduce government's telecom costs Improve local broadband infrastructure and service provision	Private sector fails to invest Bundling public telecom demand creates critical mass for broadband infrastructure
Generic supply policies	Attract businesses and citizens by improving local infrastructure	Private sector fails to invest Public company can do a better job, and will evoke more competition on service level
Supply policies in newly built/redevelopment areas	Increase the attractiveness of these areas for inhabitants and businesses	Putting fibre in the ground is relatively easy and cheap in newly built areas. It will evoke strong service competition
Group-oriented supply policies	Offer deprived groups/areas broadband access at affordable prices to stimulate social and economic development	Market fails to bring broadband to deprived groups, with negative societal consequences
Demand policies	Improve the provision of broadband infrastructure and services, with positive impact on quality of life and attracting innovative (ICT) companies to the region	Demand subsidies can resolve 'chicken and egg' deadlock that keeps companies from investing in new infrastructures

city has calculated that substantial cost savings can be achieved when the telecom expenses of all the participating organizations are capitalized and invested in the network. The city can roll out a fibre network cheaply, as it already owns two city rings, empty tubes that can relatively cheaply be filled with fibre. In the city's concept, the network will be owned by the municipality or a 100 per cent public company. Private firms will provide the services via the network. It is hoped and expected that, compared to the situation at the time of writing, more services will be offered at lower prices. At a later stage, companies and citizens are also to be connected. The community network is in an advanced stage of development: the business model is ready and there are detailed plans and cost expectations for the construction stage.

Similar initiatives are being taken in a number of cities, including Barcelona, Valencia, Leeds, Bologna and The Hague. By building community networks, cities increasingly do what private companies do not: create fibre-optic networks that connect individual buildings. Compared to private telecom companies, cities have three strengths: a great ability to bundle demand, thus creating critical mass for a high-capacity network; control over the physical infrastructure (ducts, sewage systems) that greatly facilitates the creation of such networks; and the potential to obtain relatively cheap loans.

Generic broadband supply policy

A second type of broadband policy concerns the generic supply of broadband. In this type, cities/regions roll out broadband infrastructure throughout the urban area.

Some cities roll out broadband infrastructure throughout the urban area.

The absolute European frontrunner regarding such a policy is the Stockholm municipality in Sweden. In 1994 the city decided to set up a public company, Stokab, to build a fibre network. One of the reasons for doing this was the unwillingness of the private sector (Telia, the Swedish telecom incumbent) to connect individual buildings to fibre. The city started the public company to roll out a fibre network throughout the city. The underlying vision was that a wide availability of superior broadband would make the city more attractive for inhabitants and companies. The city owns €5.5m in Stokab shares and has provided the company with €63m in loans. Stokab is an operator-independent provider that leases dark fibre connections to operators, private companies and other organizations (Isenberg, 1998). The network development began in the city's commercial district and at the time of writing covers most municipal centres and commercial areas in the region. Most of the connections encounter blocks or curbs: the 'last mile' does not yet consist of fibre. By 2002 Stokab had constructed 5,000km of fibre cable. The network developed beyond Stockholm: 26 municipalities in the Stockholm region are also connected. A total of €150m has been invested in the network. Since 1997 Stokab has been profitable and turnover has risen steeply. Some

50 operators are Stokab's clients. Most of them are service providers, but they also include private companies that prefer to have a dedicated network. An important customer of Stokab is Bredbandsbolaget: this ISP offers 10Mbps symmetrical connections to households and SMEs. In March 2001 Bredbandsbolaget had about 125,000 households connected (BDRC, 2001, p. 117).

Broadband investment in newly built/redevelopment areas
Another way local governments are involved in broadband is in newly-constructed areas. It is relatively easy and cheap to provide these areas with fibre-to-the-home. The city of Rotterdam is an example. The cities' development corporation has laid down fibre-optic infrastructure in two such areas. In total, 6,750 homes will be connected to fibre. The city will be the owner of the infrastructure. A private company is to win a concession to activate the network and other companies are invited to lease capacity and/or to offer services (TV, radio, Internet and voice) over the network. The municipality encourages competition on the network, which will hopefully result in more and better services at lower prices. In the two pilot areas, local government invested €4.6m. In the longer run, the city's ambition is to connect every house in the municipality to broadband. This would cost an estimated €200m for the passive network (Woets, 2002).

Group-oriented supply policy
As a fourth type of broadband policy, some cities seek to bring broadband to deprived groups, to tackle an emerging 'broadband divide'. Manchester provides a good example. Recently, the city of Manchester has decided to invest in a wireless local area network (WLAN) in the deprived neighbourhood of East Manchester. In an ambitious scheme, the City Council is now providing East Manchester with a WLAN. This will enable people in the area (4,500 houses) free access to the local intranet. People who want to access the Internet need to pay the normal monthly fee to an Internet Service Provider of around €24. Not only houses will be connected, but also community centres, schools and several public Internet access points. In the rollout of the network, these are the first to be connected. A private company is constructing the network. The maximum speed will be 10Mbps. This is a high speed: a typical DSL user in the UK gets a speed of 512Kbps for a monthly fee of €40.

At the time of writing, in Manchester every citizen can have dial-up Internet access using the fixed telephone network and in many places cable Internet is available. However, in practice in several deprived neighbourhoods (such as East Manchester), many people do not have a landline connection, for several reasons. Some people were simply cut off because they did not pay their bills, others switched to mobile. These people have no chance of being connected to the Internet. The urban broadband policy aims to fill this gap by offering people the opportunity to be online even when they do not have a fixed telephone connection. The municipality hopes that the WLAN access will increase social inclusion by offering the opportunity to obtain information and education, contact other people and institutions in the neighbourhood and possibly obtain a job.

Demand promotion initiatives

As a fifth type of broadband policy (a less common one), some cities are trying to promote the use of broadband Internet as well as the development of broadband by giving demand subsidies. Eindhoven is a prime example. The city has the slogan to be 'leading in technology'. It believes that an increased use of broadband can boost the local economy and attract information and knowledge-intensive companies. From this perspective, it has set up a project called Kenniswijk ('knowledge neighbourhood'): the aim is to stimulate the development of ICT broadband services and applications for end consumers. In cooperation with private enterprises and societal organizations, the city planned to create a 'pilot area' with broadband infrastructure and services, consisting of 38,000 households (84,000 inhabitants). The project started in 2001 and is subsidized by the Dutch government and Eindhoven municipality. The national government supports the project with €45.5m during a five-year period. The basic idea of the project is to break the 'chicken and egg' situation that stalls the development of the broadband market. On the demand side, the project organization provides subsidies for inhabitants who subscribe to broadband (regardless of which technology they use). For this, €13.6m is available.

The e-city organization is responsible for the management and promotes the development of broadband experiments. In the summer of 2002 a consortium of companies announced the connection of another 1,000–2,000 dwellings to fibre-optic. Individual households in the area can apply for a subsidy of €800. Of this amount, €500 is meant to subsidize the physical construction of the fibre link. This will not cover the total cost of the connection. The big question will be how many households will apply. If the number of applicants is large enough, the project will be scaled up to 15,000 households in the Kenniswijk area.

We have described a number of examples of urban broadband policies, but this number can easily be extended. Cities and regions throughout Europe are trying to promote broadband use in similar ways, often helped by national programmes and funding. In the next sections, we want to critically assess the (potential) benefits of these policies, and the associated costs/risks.

Evaluating the Benefits of Local/Regional Broadband Policies

What are the benefits for the city of connecting to broadband? Based on the policy documents and literature, at least five types of benefits can be discerned.

Urban broadband policies may add to: 1) urban attractiveness for citizens; 2) urban attractiveness for companies, 3) an improved quality or lower costs of electronic services 4) savings on public telecom costs; and 5) social equality.

Table 11.3 gives a tentative overview of the (alleged) benefits of each type of broadband policy and attaches scores to each of the type of policies discussed in the last section. The scores should be treated with caution, since no in-depth analyses were done to obtain these results and because the number of case studies is limited. Still, it is interesting to discern the differing strengths of the broadband policies. We discuss them in more detail below.

Improved urban attractiveness for citizens
Broadband Internet connections may increase the attractiveness of places to live. Recent research supports this claim. Wired homes have the potential for being seen as more upmarket and desirable than others (Baines, 2002). Hampton and Wellman (2002) found that computer-mediated communication reinforces existing communities: broadband Internet is used as a complementary way of staying in contact with neighbours. In Netville, a Toronto suburb that functions as an ICT testbed, it was found that wired residents have significantly more contact than non-wired residents. Several policy documents stress the potential of broadband for improving the quality of services and amenities. According to Leighton (2001) broadband will significantly change and improve peoples' daily lives: it gives them improved opportunities to receive education in their own homes, intricate medical data can be swiftly sent within and between medical institutions and telecommuting, business conferences and personal communications will benefit from using broadband. In an advice to the government, the Dutch Expertgroup Broadband considers broadband access as the oxygen of society (Expertgroep Breedband, 2002). From a societal perspective, it argues, broadband availability can have positive impacts on education, governance, safety, culture, trade and leisure.

All the policy types have a positive score in this category, although the number of citizens they reach differs. Stockholm scores highest, because its network encompasses the entire region and offers most broadband connections. The community network scores a 0, as we have assumed that only (semi-)public actors are connected. However, if such networks are extended to companies and citizens, the score becomes positive too.

Improved urban attractiveness for companies
Electronic infrastructure is increasingly recognized as an important location factor for companies, especially knowledge- and information-intensive ones (Healey and Baker Consultants, 2001). The Dutch 'Expertgroep Breedband' links the growth of broadband Internet to the positive growth of the ICT sector in The Netherlands, and argues that 50 per cent of the productivity increase can be ascribed to ICTs, in which broadband plays a prominent role (Expertgroep Breedband, 2002).

Electronic infrastructure is increasingly recognized as an important location factor for companies, especially knowledge and information intensive ones.

Table 11.3 Benefits of various types of broadband policies

Contribution to ...	Urban attractiveness for citizens	Urban attractiveness for companies	Quality of electronic services/ lower prices	Public telecom costs savings	Social equality: equal broadband access
Type of policy:					
Community networks	0	0	+	++	0/+
Supply policies in newly built/ redevelopment areas	+	+	+	0	0
Demand policies	+	+	0/+	0	0
Generic supply policies	++	++	++	+	0
Group-oriented supply policies	+	0	0	0	+

The cities and regions in our sample all regard the supply of broadband as a location factor: they hope to become more attractive to companies and citizens by offering better infrastructure and connections. Apart from general benefits, cities and regions perceive benefits from being (one of) the first to have high levels of broadband adoption, or 'first mover advantages'. Eindhoven, for instance, hopes to become a 'broadband laboratory' that attracts innovative service providers and other businesses that are interested in many citizens connected to broadband Internet. As can be seen from Table 11.3, three of the five policy types will probably increase the attractiveness for companies. In each of them, companies get better and/or cheaper Internet access. The community networks get a 0 score, as no companies are connected, and the group-oriented policies (Manchester) are only directed to individuals and households.

Improved quality of electronic services/lower prices
In the current situation, many cities have just one or two broadband Internet suppliers (typically, these are the incumbent telecom company that offers DSL on the copper telephone network and the broadcasting company that offers cable Internet). Even if cities have more suppliers, these parties are dependent on the networks of the incumbents since they don't own their own networks. The incumbent players try hard to impede the delivery of competitors' services, often with success. If local governments invest in new infrastructures, notably in fibre-optic, the competition in the broadband market can be stimulated, if governments decide to give the network ownership a public character and offer multiple service providers the possibility to use the network simultaneously. This can lead to lower prices and improved services. From this perspective, the generic supply policy of Stockholm has the highest score; Manchester's digital divide initiative has no impact, as there will be no service competition.

If local governments invest in new infrastructures, notably in fibre-optic, the competition in the broadband market can be stimulated, if governments decide to give the network ownership a public character and offer multiple service providers the possibility of using the network simultaneously. This can lead to lower prices and improved services.

Public telecom costs savings
An explicit motive by some cities (for example Groningen) is to achieve telecom costs savings by creating a broadband community network that will handle all data traffic (including internal telephone traffic) from the organizations connected. The community network will replace all the telephone and data connections that were first bought from private companies. The cost savings can be used to lower municipal taxes or increase service delivery to citizens. Cities also hope to promote competition for services on the new public fibre-optic networks. As can be seen from Table 11.3, the community networks have the highest score in this category.

Modest positive impacts can be expected from generic supply policies (Stockholm). The impact of the other policy types is negligible.

Increased social equality; equal access to broadband
Lack of access to broadband limits the possibilities of accessing the Internet. This limited access to information can be considered as a form of social exclusion (Castells, 1998; Graham, 2002), which can be prevented/fought by giving people access to broadband. Many cities stress the importance of broadband as a means of obtaining ICT skills, strengthening inhabitants' knowledge, extending social contacts and increasing job opportunities. Manchester primarily sees social benefits in its WLAN initiative: it is hoped that connecting people will give them more self-esteem, facilitate their way back to the labour market and promote social inclusion through the formation of lively online communities. In Table 5.1, Manchester's broadband policy gets the highest score, as it explicitly addresses slow-adopting groups. The community network policy has some potential, as many cities explicitly state that they see these networks as a tool for 'e-inclusion' (by setting up public Internet points in schools, libraries, municipal offices, etc).

Costs and Risks of Local/Regional Broadband Policies

The potential benefits of the broadband policies do not justify the policies *per se*. As in any evaluation, they should be critically assessed and weighed against the costs and risks of the policies. Also, the question should be raised of whether an alternative allocation of (public) means could yield similar or better results: ICT investment should never be a goal in itself but a means to achieve economic development, social inclusion or other policy goals.

Many proponents of government intervention argue that the market outcome is undesirable: in their view, broadband adoption rates remain too low and weaker groups are largely excluded. This would justify policy measures. But to what extent are low adoption rates a problem? The adoption of a new technology typically follows an S-shaped curve. In the first stage, adoption only grows slowly. Rich and/or highly educated people are the early adopters. From a certain point, the number of users grows very quickly, as positive network externalities become stronger and prices fall. When the adoption rate reaches maturity, growth slows down again (Shapiro and Varian, 1999). This pattern can be observed for the adoption of electricity, telephone, television, cable television networks, personal computers and the Internet. See Figure 11.4 for a number of S-curves.

Geographically, metropolitan areas are the first places where new technologies are available, and small towns, rural and remote areas follow later. Interestingly, with new technologies S-curves seem to become steeper: the time span between the introduction of the new technology and adoption by the majority has decreased dramatically. This observation puts policies to fight the 'digital divide' or 'broadband divide' into question.

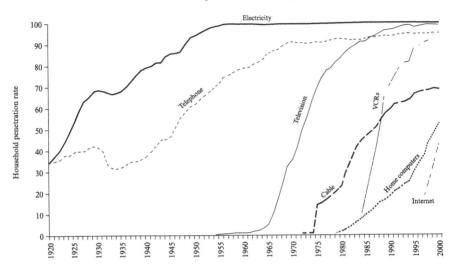

Figure 11.4 Technology adoption S-curves in the US

Source: Leighton (2001).

> *The issue should not be to fight a digital divide (either social or spatial), but to fight it where the divide is persisting and damaging.*

The issue should not be to fight a digital divide (either social or spatial), but to fight it where the divide is persisting and damaging. Concerning broadband, the typical S-shape curve seems to apply. Adoption rates, although still low in many countries, are rising quickly and prices continue to fall. From this perspective, policy can best be targeted at two categories: weak social groups (as in Manchester) and people in remote areas. For both groups, the reasons for low adoption are very different, however. For weaker social groups, low adoption rates stem from low incomes, low educational levels, weak social support and high unemployment (van Winden, 2001; SCP, 2000). Van Winden (2001) argues that the digital divide is a mere derivative of these factors and concludes that generic social inclusion and labour market policies are indirectly more likely to promote ICT adoption than directly subsidizing ICT use.

In addition, there are two costs/risks that should be taken into account concerning broadband policy – costs to taxpayers and unbalancing a competitive industry.

Costs to taxpayers
The costs of many broadband policies are ultimately carried by the taxpayer. Policymakers should realize that these funds are not available for other policy objectives, or that they should be raised with higher taxes. The burden is spread across all taxpayers, while the benefits accrue only to specific groups. In our cases,

costs vary substantially. The costs of the policies are summarized in Table 11.4. They are stated in terms of investments. Per connection, investments are the highest in Eindhoven and substantially lower in the other cases. The Eindhoven investments, however, include infrastructure, access and services, while the investments in the other cities only involve infrastructure. The WLAN network in East Manchester is estimated to cost €3.2m and will connect 4,500 houses (van der Meer and van Winden, 2002), which implies €710 per connection. Much of this investment is carried by either national or European funds.

Table 11.4 Costs and potential returns of broadband policies

Case	Total investment (€)	Investment per connection (€)	Return on capital
Eindhoven	45 m	1,185	None
Rotterdam	4.6 m	680	Potentially positive
Manchester	3.2 m	710	None
Stockholm	150 m	?	Positive

Note: The number of broadband connections for Stockholm is unknown; this implies that the investment per connection could not be calculated.

For Rotterdam and Stockholm, there are (potential) returns on capital as the passive infrastructure is to be leased out at commercial rates. Stokab is currently generating profits that come back to the municipality. The investments in Eindhoven are to a large extent provided through national funds.

Unbalancing a competitive industry
The cities of Manchester and Rotterdam each clearly support one technology: fibre-optic technology and WLAN technology respectively. However, Leighton (2001) argues that policies can distort competition and disturb competitive relations if they clearly benefit one or a few companies, or favour one particular technology. According to Umino (2002), broadband development is so dynamic and in such an embryonic stage that it is too early for the government to implicitly exclude other possible technologies. Others (Eindhoven) claim that their policies are 'technology neutral' and do not favour any technology, infrastructure or service provider over another. Leighton (2001), however, argues that even in the case of seemingly technological neutrality, existing dominant suppliers of broadband services are likely to benefit most, as well as suppliers of equipment to activate the infrastructures. This result is not neutral, as these technologies will be challenged by new ones. New competitors will emerge if they can challenge existing companies with lower prices or better service, or when they can enter markets that are not

served yet. Thus, 'government programs that benefit existing providers ultimately reduce incentives to develop advances in service' (Leighton, 2001).

In the cases of Groningen, Rotterdam and Stockholm, the city itself (in the form of a public company) becomes a player in the local telecom market, as owner or major shareholder of the fibre-optic infrastructure. Umino (2002) warns that this may lead to the creation of local monopolies. As a result, in the future, commercial players are less likely to invest in infrastructure in that area, even if better technologies are available by that time. The infrastructure market is distorted, although service competition on the infrastructure will probably increase. Umino (2002) mentions several other drawbacks of public ownership: 1) it is costly to maintain the network and upgrade it as technology changes; 2) public work projects are prone to inefficiency; and 3) as a network owner, the government is badly placed to separate its regulatory role from its role as a network supplier. This may lead to unfair competition.

In the longer run, no one knows what technologies will best provide broadband. If existing providers benefit from broadband promotion policies, this creates a disincentive for new technologies. Tax credit or subsidy policies (as in Eindhoven) enable existing providers to offer their old technologies at lower costs, or extend their market to geographical areas they would not service without such policy. For newcomers it is less attractive/profitable to enter a new market when existing providers already offer broadband. In the end, broadband policies may have the effect that newer and better technologies do not get a fair chance.

Conclusions

In this section, we described various forms of local infrastructure policies. We have confronted the potential benefits of the various policy types with an analysis of the costs and risks of associated policies. Based on this analysis, we draw a number of conclusions.

One option for cities is to do nothing and let the market do its job. Like any technology, DSL or cable Internet take-up will increase and, later, private companies may be willing to invest in fibre or other advanced infrastructure to bring even higher speed access. This option is cheap, but it may take a long time before broadband is available to large numbers of consumers and companies. This may undermine the competitiveness and productivity of companies and people in those cities. Demand policies (as in Eindhoven) have the merit of distorting the market only slightly. However, they do not seem to bring the expected investments in broadband infrastructure. In the current development stage of the broadband market, and the situation of the incumbents (that dominate the DSL market), large scale investments in broadband infrastructure only come with heavy government support.

When the benefits, costs and risks are weighed, the most successful type of policy seems to be the generic supply policy in the Stockholm case, where the city has managed to build an extensive fibre infrastructure and increase competition on services level. The strict separation of infrastructure ownership and services has been a key factor. The supply model of Rotterdam (constructing fibre in newly-built areas) and the community network (Chicago, Groningen) may develop into a similar structure in the long run. The counter-argument that such broadband policy would 'thwart a competitive industry' is questionable: in most regions the broadband market is not that competitive, since telecom and television cable incumbents dominate it. On top of that, these types of local broadband policy may well lead to more competition instead of less, when services competition is set to increase. Opponents argue that governments should not promote one technology as, given the rapid technological developments, no one knows what the future will bring. Although in theory this is true, fibre-optic is widely regarded to as a long-term future proof broadband conveyor: it is used as a backbone infrastructure for all the other broadband technologies.

Broadband policy to fight a digital divide is questionable from a 'dynamic adoption' perspective (S-curve), and in the case of Manchester is also quite costly. Where social inclusion is aimed for, an alternative allocation of means is likely to yield higher returns.

Cities invest a lot of money in broadband and in the near future many more millions are likely to be invested. This study hopefully contributes to this discussion. However, many questions remain still unanswered. First, broadband claims to bring a lot of benefits, but in-depth studies are needed to find out to what extent and under what conditions broadband really contributes to economic development, social inclusion and a more efficient government. Another key question concerning urban development is how the broadband adoption will affect location decisions of individuals/firms and consequently spatial patterns (living/working), and whether urban regions will extent their competitive advantage over remoter towns through higher broadband deployment.

Recommendations and Policy Lessons

This report has made an attempt to describe and analyse 'e-strategies' of a number of European and South African cities. In this last chapter, we have synthesized the findings. In this section, we want to draw lessons and give policy recommendations.

It can be expected that the full impact of ICTs on urban change has not materialized yet, given the continuing rapid developments in ICTs and the long time that passes before a new technology (or set of technologies) fundamentally changes social and economic processes. In much of the Western world (and increasingly in the developing world, too), access to ICTs and electronic infrastructure is increasingly well developed. Physical electronic infrastructure is ubiquitous (although large spatial differences in quality and price exist, as we have argued in this report) and an increasing percentage of the population has ICT skills and access to personal computers and networks. In contrast, the evolution on the content side – in the widest sense of the word – is much slower. Governments and companies have great difficulties in effectively using new ICTs and still tend to stick to existing organizational structures and routines. Governments also have great difficulties in transforming their work processes and using technology to become more client-oriented, responsive and efficient. Companies are only just beginning to reshape their organizations based on the logic of networks, although some sectors are more advanced than others in this respect. The financial sector has already fundamentally changed under influence of ICT adoption and will continue to do so. The media and entertainment industries are currently in a process of transformation too, and in other sectors a radical adoption of network technologies also brings fundamental changes. Easyjet and Ryanair in the airline industry and Dell computers in the computer industry are good examples: in a very short time, they gained large market share with a business model based on the new possibilities of electronic networks.

There is a time lag between the development access and infrastructure on the one hand, and the transformations that come with content on the other. An analogy may be drawn with the spatial economic impacts of transport infrastructure, where there is a substantial time lag between the construction of new transport infrastructure and the spatial redistribution of activities that takes place afterwards. If we transplant this notion to the urban level, it can be argued that the major urban impacts of ICTs are yet to come.

From this perspective, local governments have two major challenges. The first is to understand and monitor the global social and economic transformation processes caused by ICTs in combination with other factors, to estimate their impact on the specific urban context and to take adequate policy measures to achieve sustainable urban development. In a general sense, for cities this implies creating conditions for entrepreneurship and creativity; investing in culture, events, education and amenities; trying to attract talent to the city; making most out of the city's knowledge- and creativity base, and ensuring equal opportunities for all, not only to prevent segregation and social unrest but also not to waste talent.

The second challenge for cities is to adopt an adequate technology management which balances content, access and infrastructure. In the last decade, as this report shows, local governments have been busy with e-access, ensuring everyone has access to new technologies and recently, broadband infrastructures have received much policy attention. In the near future it will be time for content: the more complex step will be to put the technology into use to effectively improve the quality of the city and the city management. The possibilities are endless, but the realization is difficult and asks for new approaches. Challenges will be, for example, to use mobile technologies for tourism, to achieve integrated city marketing using web technology and to apply ICTs in the public and semi-public sector in order to make these sectors more efficient and effective. This requires the ability to understand technological possibilities and limitations, and organizing capacity to bring urban actors, technology suppliers and telecom companies together and create new combinations. The traditional municipal hierarchies are poorly suited to manage technology projects as they tend to be insufficiently demand oriented, risk averse and unresponsive. Furthermore, in defining technology projects, a shift from a supply-oriented 'inside out' perspective to a demand-led 'outside in' approach is needed. This is a radically different from current practices in many cities, where projects are too often defined on the basis of the possibilities of new technologies instead of the actual need of city users and an analysis of the contribution that ICT projects could make to sustainable urban development.

Moreover, this report shows that cities far from optimally involve the private sector: they are only beginning to develop public private co-operations in ICT policy. For urban technology management (but also in other policy fields), a major issue is how to make optimum use of private sector expertise, funds and responsiveness, without compromising accountability and democratic control.

The two challenges mentioned above are interrelated: an adequate urban technology management can contribute to create conditions for sustainable urban development. ICT is one tool – alongside many others – to improve the quality of life and accessibility, to empower citizens, to improve decision-making processes and to enhance democracy.

For Further Research

This report has filled a gap in the existing literature on urban development, by developing an analytical tool to make a structural analysis of urban ICT policies. However, a lot of research work still needs to be done in the fields of ICT and urban development. Here, we list a number research topic and questions.

- *An empirical evaluation of the 'local digital flywheel'.* In Chapter 2 we presented the 'digital flywheel', in which we assumed that access, content and infrastructure are mutually reinforcing and could have a positive impact on the various dimensions of urban development. Although some parts of that conceptual model have been further elaborated in this report, the model could be tested much more extensively. One option is to design a measure for the quality of local content, a measure for levels of access and a measure for the quality of local infrastructure. An analysis of these variables for a representative set of cities could reveal whether the three aspects are indeed related. To research the assumed positive relationship with the various dimensions of urban development (economic growth, social inclusion, accessibility and quality of life) it could also be tried to link these values to urban performance indicators in these dimensions.
- *The effectiveness and efficiency of e-government.* During the 1990s, ICT spending was justified by the general belief that ICT would make governments better: it would streamline internal processes, increase efficiency, reduce costs, improve services delivery and enhance the quality and speed of decision-making. Paradoxically, citizens throughout Europe seem to be less happy with the quality of their governments than a decade ago. This raises the question

how much has ICT really achieved. A longitudinal analysis of the quality of public service delivery could reveal this. Relevant questions could be, to what extent has ICT led to substantial improvements in public service delivery? To what extent has it reduced costs and improved efficiency? What success factors can be found? How can mistakes be avoided?

- *The urban economic impact of local broadband policies.* Cities invest a lot of money and resources in broadband infrastructures and in the near future many more millions are likely to be invested. This requires more in-depth research into the economic and social impacts of broadband. First, broadband claims to bring a lot of benefits, but in-depth studies are needed to find out to what extent and under what conditions broadband really contributes to economic development, social inclusion and quality of life variables. Another key question concerning urban development is how broadband adoption will affect location decisions of individuals/firms and consequently spatial patterns (living/working), and whether urban regions will extent their competitive advantage over remoter towns.
- *The effects of ICT on social inclusion.* Much additional empirical research is needed to improve our understanding of the role of ICT in social inclusion and exclusion. Processes of social exclusion are complex and far from completely understood; also, relatively little is known about the way ICT affects people's behaviour. Thorough in-depth empirical research is needed to study the uptake of ICTs by weaker social groups, and the difference it makes to their participation in social, economic and political life.
- *ICT and location behaviour.* To understand the spatial-economic consequences of ICT adoption, we need a good conceptualization of the way ICT impacts/ alters the location behaviour of companies, citizens and other actors.

References

Algemeen Dagblad (2002), 'Digistroom: daad of droom?', 6 November.

Baines, S. (2002), 'Wired Cities', *Communications International*, April, pp. 21–25.

BDRC Ltd (2001), *The Development of Broadband Access Platforms in Europe*, London: http:www/europa.eu.int/eeurope.

Castells, M. (1998), *End of Millennium*, Cambridge, MA: Blackwell Publishers.

DTI (2000), 'Closing the Digital Divide: Information and Communication Technologies in Deprived Areas', a report by Policy Action Team 15, Department of Trade and Industry, London.

DTLR (2002), *E-gov@local: Towards a National Strategy for Local E-government*, Department of Transport, Local Government and the Regions, London.

Expertgroep Breedband (2002), 'Nederland Breedbandland – advies aan het kabinet', The Hague.

Graham, S. (1998), 'The End of Geography or the Explosion of Space? Conceptualising Space, Place and Information Technology', *Progress in Human Geography* 2, pp. 165–85.

Graham, S. (2002), 'Bridging Urban Digital Divides? Urban Polarisation and Information and Communications Technologies (ICTs)', *Urban Studies* 39 (1), pp. 33–56.

Hall, P. (1998), *Cities in Civilization: Culture, Innovation and Urban Order*, Weidenfield and Nicolson, London.

Hampton, K.N. and Wellman, B. (2002), 'The Not So Global Village of Netville', in B. Wellman and C. Haythornthwaite (eds), *The Internet and Everyday Life*, Oxford: Blackwell, pp. 345–71.

Healey and Baker Consultants (2001), *European E-locations Monitor*.

Isenberg, D. (1998) 'A Tale of Two Cities', *America's Network* 102 (19): http://www.americasnetwork.com/.

Leighton, W.A. (2001), *Broadband Deployment and the Digital Divide: A Primer*, Policy Analysis No. 410, August 7.

Mitchell, W.J. (1999), 'Equitable Access to the Online World, in High Technology and Low-income Communities', in D.A.B. Schön, A.B. Sanyal and W.J. Mitchell (eds), *Prospects for the Positive Use of Advanced Information Technology*, MIT Press, Cambridge.

Mom, P. (2003), 'Superpilots zerzanden in eigen organisatiekwesties', *Overheid Innovatief*, 3, pp. 34–7.

NOM (Northern Netherlands Development Company) (2002), *Stimulering Breedband-technologie in Noord-Nederland*, Groningen.

SCP (2000), *Digitalisering van de leefwereld: een onderzoek naar informatie- en communicatietechnologie en sociale ongelijkheid*, Sociaal en Cultureel Planbureau, Den Haag.

Shapiro, C. and Varian, H.R. (1999), *Information Rules*, Boston, MA: Harvard Business School Press.

Symonds, M. (2000), 'Government and the Internet: Haves and Have-nots', *The Economist*, 24 June.

Umino, A. (2002), 'Broadband Infrastructure Deployment: The Role of Government Assistance', OECD Working Paper No. 15.

van den Berg, L. and van Winden, W. (2002), *Information and Communication Technology as Potential Catalyst for Sustainable Urban Development*, Euricur Series, Ashgate, Aldershot.

van den Berg, L., Braun, E. and van der Meer, J. (1997), 'The Organising Capacity of Metropolitan Regions', *Environment and Planning C: Government and Policy* 15, pp. 253–72.

van der Meer, A. and van Winden, W. (2003), 'E-governance in Cities: A Comparison of Urban Information and Communication Technology Policies', *Regional Studies* 37 (4), pp. 407–19.

van Winden, W. (2001), 'The End of Social Exclusion? On Information Technology Policy as a Key to Social Inclusion in Large European Cities', *Regional Studies* 35 (9), pp. 861–77.

Woets, P. (2002), 'The Role of Governments in Providing Fibre-to-the-Home', MSc thesis, Erasmus University Rotterdam.

Index

Note: numbers in bold represent figures and tables.

Printed in the United States
by Baker & Taylor Publisher Services